Business Rule-Oriented
Conceptual Modeling

W0051415

Contributions to Management Science

Ulrich A. W. Tetzlaff
Optimal Design of Flexible Manufacturing Systems
1990, 190 pp. ISBN 3-7908-0516-5

Fred von Gunten
Competition in the Swiss Plastics Manufacturing Industry
1991, 408 pp. ISBN 3-7908-0541-6

Harald Dyckhoff/Ute Finke
Cutting and Packing in Production and Distribution
1992, 248 pp. ISBN 3-7908-0630-7

Hagen K. C. Pfeiffer
The Diffusion of Electronic Data Interchange
1992, 257 pp. ISBN 3-7908-0631-5

Evert Jan Stokking/Giovanni Zambruno (Eds.)
Recent Research in Financial Modelling
1993, 174 pp. ISBN 3-7908-0683-8

Richard Flavell (Ed.)
Modelling Reality and Personal Modelling
1993, 407 pp. ISBN 3-7908-0682-X

Lorenzo Peccati/Matti Virén (Eds.)
Financial Modelling
1994, 364 pp. ISBN 3-7908-0765-6

Michael Hofmann/Monika List (Eds.)
Psychoanalysis and Management
1994, 392 pp. ISBN 3-7908-0795-8

Rita L. D'Ecclesia/Stavros A. Zenios (Eds.)
Operations Research Models in Quantitative Finance
1994, 364 pp. ISBN 3-7908-0803-2

Mario S. Catalani/Giuseppe F. Clerico
Decision Making Structures
1996, 175 pp. ISBN 3-7908-0895-4

M. Bertocchi/E. Cavalli/S. Komlósi (Eds.)
Modelling Techniques for Financial Markets and Bank Management
1996, 296 pp. ISBN 3-7908-0928-4

Holger Herbst

Business Rule-Oriented Conceptual Modeling

With 108 Figures

Springer-Verlag Berlin Heidelberg GmbH

Series Editors
Werner A. Müller
Peter Schuster

Author
Dr. Holger Herbst
Institute of Information Systems
Research Group 'Information Engineering'
University of Bern
Engehaldenstr. 8
CH-3012 Bern, Switzerland

ISBN 978-3-7908-1004-2

Cataloging-in-Publication Data applied for
Die Deutsche Bibliothek – CIP-Einheitsaufnahme
Herbst, Holger: Business rule-oriented conceptual modeling / Holger Herbst. – Heidelberg: Physica-Verl., 1997
   (Contributions to management science)
   ISBN 978-3-7908-1004-2   ISBN 978-3-642-59260-7 (eBook)
   DOI 10.1007/978-3-642-59260-7

Softcover Design: Erich Kirchner, Heidelberg

SPIN 10569852   88/2202-5 4 3 2 1 0 – Printed on acid-free paper to Bea

PREFACE

This book is based on a PhD dissertation which was accepted by the faculty of Law and Economics at the University of Bern, Switzerland. The ideas presented were partially developed in a research project founded by the Swiss National Science Foundation in 1993 and 1994. This research project was concerned with evaluating the application of database triggers and active databases for the implementation of business rules. We recognized among other things the lack of a methodology for modeling such business rules on the conceptual level. Therefore, this became the focus of the follow-up research which resulted in this book.

All this work would not have been possible without the help of several people. First of all, I would like to give special thanks to my thesis supervisor Prof. Dr. Gerhard Knolmayer. He not only initiated the research project and found an industry partner, but also provided very valuable ideas, and critically reviewed and discussed the resulting publications. Furthermore, I would like to express my thanks to the second thesis supervisor Prof. Dr. Sham Navathe from Georgia Institute of Technology who influenced my work with results from a former research project and who agreed to evaluate the resulting PhD Dissertation.

I am also grateful to all my colleagues of the Institute of Information Systems, namely Dr. Thomas Myrach, Dipl. Inf. Markus Schlesinger, and Dr. Beat Jaccottet for their ideas and critical discussions. Furthermore, I would like to mention lic. rer. pol. Petra Suter, lic. rer. pol. Patrick Eisenegger, and lic. rer. pol. Mathias Rytz whose master theses contributed to this book.

Beside all these people from academia, I am obliged to the „Schweizerische Mobiliarversicherungsgesellschaft", which served as industrial partner of my research and provided my with valuable insights in its business processes and the underlying information systems. Finally, I'd like to thank everyone who gave me critical feedback by reviewing my papers or discussing my ideas at the conferences where I was allowed to present my work.

Bern, November 1996   Holger Herbst

TABLE OF CONTENTS

TABLE OF FIGURES

1 INTRODUCTION

Since the industrial revolution manifold organizational approaches have been developed and more or less successfully applied in practice. One of the most recent trends of organizational reflections is *business process reengineering*[1]. Within all these approaches the use of formalized rules is a widely discussed and controversial topic. Most organizations today make an intense use of information technology (IT) in order to support their functions and processes and to finally achieve good positions in a competition which is becoming more and more intensified[2]. The information systems (IS) as a part of IT encompass a huge amount of often heterogeneously structured and formalized organizational knowledge. This knowledge is often referred to as *business rules*[3] and many business processes are executed with respect to those rules which either prescribe a certain action or constrain the set of possible actions. In recent years, it has been accentuated that business rules are an important element of many IS; even the notion of a paradigm change has been used in emphasizing their importance for conceptual modeling[4]. Loucopoulos demands that „all laws and rules governing the universe of discourse must be defined within the conceptual schema."[5] Though the term *business rule* is often referred to, it is defined and used rather differently and often confined to semantic integrity constraints[6]. However, business rules do not only cover data integrity but may also impose restrictions on organizational dynamics; therefore, with reference to Bell et al.[7], we define business rules as

> statements about how the business is done, i.e., about guidelines and restrictions with respect to states and processes in an organization.

For the development of IS, different methodologies have been proposed and are partly used in practice. A basic paradigm for the modeling of IS is the separation

---

[1] Cf. e.g. Hammer/Champy(1993); Davenport (1993).
[2] Cf. Hammer/Champy (1993), p. 21.
[3] Cf. e.g. Appleton (1995), pp. 291; Hammer/Champy (1993), pp. 42.
[4] Cf. Van Assche et al. (1988); Moriarty (1993a).
[5] Loucopoulos (1992), p. 6.
[6] Cf. e.g. Appleton (1984); Appleton (1988); Sandifer/VonHalle (1991); Moriarty (1993b); Von Halle (1993a).
[7] Cf. Bell et al. (1990a); Bell et al. (1990b).

of three abstraction levels known as the ANSI/SPARC 3-level-architecture[8]. The core idea of this architecture is to separate models of the external, the conceptual, and the physical layer:

- The *external model* represents the view of users or user groups on the IS.
- The *conceptual model* is an ideal model which describes the universe of discourse without considering any implementation aspects.
- The *physical model* depicts the technical implementation of the conceptual model.

This leveled modeling architecture leads to an independency between models of different levels, i.e., any modifications in the external model (e.g. to add the view of a new user group) or the physical model (e.g. in order to optimize the performance) can be done without changing the conceptual model. The conceptual model is an abstract representation of the real world phenomena relevant for an IS and serves as[9]

- a common reference framework which is used to communicate with current and future end users,
- a model of reality which enables analysts to understand the concerned real world,
- a basis for the design and implementation of an IS, and
- a part of the documentation which can be used during the maintenance of the IS.

Today's enterprises mostly make use of database management systems (DBMS) for the storing and administration of information. Since the introduction of the first DBMS[10] the database research has resulted in different data models[11]: the hierarchical, the network, the relational, and most recently the object-oriented model. An important direction of current DBMS research attempts to integrate *rules* in relational or object-oriented DBMS[12]. As a result of this research, prototypes of active DBMS[13], e.g., SAMOS[14], are becoming available. Furthermore, several commercially available DBMS provide rule mechanisms, e.g., database

---

[8] Cf. ANSI/X3/SPARC (1975).

[9] Cf. Kung/Sølvberg (1986), pp. 146.

[10] For an historical overview of DBMS cf. Elmasri/Navathe (1994), pp. 20; Küng (1994a), pp. 58.

[11] A comprehensive introduction to the different data models can be found in Elmasri/Navathe (1994).

[12] Cf. the discussion of active DBMS in Section 2.2.2.3.

[13] Cf. e.g. Morgenstern (1983); Dittrich et al. (1995).

[14] Cf. Gatziu et al. (1994).

triggers which, however, have a limited functionality only[15]. Rules in active DBMS often consist of the three components *event*, *condition*, and *action* and are called ECA rules[16]: the event component indicates *when a rule has to be executed*, the condition *what has to be checked*, and the action *what has to be done*.

The use of active rules for the implementation of IS may lead to very large and therefore complex rule systems which may become impossible to administer. Furthermore, because the action of a rule may trigger other rules, the behavior of the whole IS may become unpredictable. A similar situation existed several years ago when complex situations in the area of data storage in DBMS resulted in the need for conceptual data modeling[17]. Therefore, all the developments on the physical level „suggest the need for a systems analysis technique that captures not just data, process and dynamics but rules and facts of the type found in knowledge based systems. Object-oriented databases and extended [e.g. active] relational databases will also make it imperative that business rules are captured explicitly during the analysis process.“[18] However, the modeling of such rules on the conceptual level has so far been widely neglected[19], though the very simple ECA structure provides a huge potential to specify behavioral knowledge[20].

The ECA structure allows for specifying integrity constraints as well as business processes consisting of business rules as main formalization elements. As an example of an integrity constraint formalized as ECA rule, the following rule prevents a stock from falling below zero:

> *ON   (stock reduced)*
> *IF   (new stock < 0)*
> *THEN   reject operation; error message „the stock must not be below zero“*

The specification of business processes by using ECA rules is possible, because actions executed by one rule may lead to events which trigger other rules, thus describing the behavior of processes. The following example describes a simple stock reorder rule:

> *ON   (stock reduced)*
> *IF   (new stock < reorder threshold)*
> *THEN   reorder product*

---

[15] Cf. Schlesinger (1995).

[16] Cf. Dayal (1988).

[17] For an overview of semantic database modeling cf. e.g. Hull/King (1987).

[18] Graham (1995), p. 58.

[19] In Chang/Chang (1982) the use of ECA rules as alerters in office automation is discussed.

[20] In Tsalgatidou et al. (1990), pp. 254, a kind of ECA rules is used to model dynamic aspects of IS.

The execution of the action *reorder product* can be regarded as an event *product reordered* which may trigger another rule:

> *ON*   (*product reordered*)
> *IF*   (*supplier out of stock*)
> *THEN*   *inform sales department*

The possibility of specifying processes using ECA rules, makes these rules applicable e.g. within concepts of business process reengineering[21] or workflow management[22]. Especially the importance of events with respect to processes is emphasized[23] and there exist several approaches which focus on the use of events[24]. The use of conditions as *post*conditions of events, however, is a rather new way of modeling. As stated by Olle et al., some „event modelling techniques use conditions in both roles (namely pre- and post-) and others use only preconditions. Expressing postconditions on BUSINESS EVENTS is not normally allowed in most methodologies"[25]. In Färberböck et al.[26], events on the conceptual level are considered as a major improvement since the first introduction of structured analysis.

The main goal of our research project was to investigate the applicability of ECA rules for the formalization of business rules on the conceptual level. An early result of our case studies[27] in which business rules were extracted from practically applied IS was the extension of the basic ECA structure to ECA^2 (ECAA) [28]:

- *Event*: When has a business rule to be processed?
- *Condition*: What has to be checked?
- *Then-Action*: What has to be done if the condition is **true**?
- *Else-Action*: What has to be done if the condition is **false**?

This extended structure allows for a less redundant specification of business rules, provides a selection between two alternative actions, and can always be transformed back into two basic ECA rules in which the condition has to be specified redundantly (once negated):

---

[21] Cf. e.g. Hammer/Champy (1993); Davenport (1993).

[22] Cf. e.g. Heilmann (1994); Kappel et al. (1994c).

[23] Cf. e.g. Perry/Denna (1995); Denna/Perry (1995); Streng (1994).

[24] Cf. e.g. Bubenko (1980); De Antonellis/Zonta (1981); Rolland/Richard (1982); Rolland et al. (1988); Österle/Gutzwiller (1992a); Scheer (1995); Dubois et al. (1986); Souveyet/Rolland (1990); Teisseire et al. (1994); Belkhatir/Melo (1994); Iivari/Koskela (1993); Bertocchi et al. (1993); Neuhold (1983).

[25] Olle et al. (1991), p. 91.

[26] Cf. Färberböck et al. (1991), pp. 60.

[27] Cf. Chapter 3.

[28] In the following, the acronym ECA^2 will be used for ECAA.

$$
\begin{array}{ll}
ON & event_1 \\
IF & condition_1 \\
THEN & action_1 \\
ELSE & action_2
\end{array}
$$

$$
\begin{array}{ll}
ON & event_1 \\
IF & condition_1 \\
THEN & action_1
\end{array}
$$

$$
\begin{array}{ll}
ON & event_1 \\
IF & NOT\ condition_1 \\
THEN & action_2
\end{array}
$$

As indicated above, business rules may be used to specify integrity constraints as well as processes. In most current approaches the formalization of these two aspects is done with different modeling constructs. However, the distinction may be ambiguous, because the result of the evaluation of an integrity constraint may be relevant within a business process and may trigger further process steps. Therefore, the rather artificial separation of the specification may lead to a gap in the process.

The relevance of business rules in business processes[29] is exemplified by Hammer and Champy:

„The reengineering process at Ford breaks hard and fast rules that formerly applied there. Every business has these rules, deeply ingrained in the operation of the organization, whether they are explicitly spelled or not.

For instance, Rule One at Ford's accounts payable department was: We pay when we receive the invoice. While this rule was rarely articulated, it was the frame around which the old process was formed. When Ford's managers reinvented this process, they were effectively asking whether they still wanted to live by this rule. The answer was no. The way to break this rule was to eliminate invoices. Instead of 'We pay when we receive the invoice,' the new rule at Ford is 'We pay when we receive the goods.' "[30]

Using the structure of business rules these two rules would be formulated as follows:

Rule One (old)
$$
\begin{array}{ll}
ON & (invoice\ received) \\
THEN & pay\ bill
\end{array}
$$
⇩
Rule One (new)
$$
\begin{array}{ll}
ON & (goods\ received) \\
THEN & pay\ bill
\end{array}
$$

---

[29] In Franckson (1994), p. 145, the Euromethod framework is described in which business rules govern business processes.

[30] Hammer/Champy (1993), pp. 42.

Only a small subset of the enormous number of business rules relevant for an organization is enforced by application programs of the IS. Those rules implemented and processed by the IS are often scattered all over the system. This may lead to incompleteness and redundancy of these rules because they may be specified in a different way in several applications and/or may be implemented differently. Even if they are implemented consistently, this property may get lost by uncoordinated maintenance in redundant pieces of code[31].

Summarizing the discussion above, there are strong arguments for a modeling of business rules on the conceptual level[32]. Therefore, we investigated the properties of a business rule-oriented conceptual modeling (*BROCOM*) based on the following assumptions:

- Business rules are an important element of organizations and have to be emphasized in the conceptual modeling of IS[33].
- Business rules can be described using the ECA^2 structure.
- Repository systems provide powerful functionality to administer large quantities of meta data[34] which is to be exploited for *BROCOM*.
- There exists an immense number of graphical models for the representation of different views on the universe of discourse. Therefore, *no new graphical model* is to be developed because there should exist at least one appropriate representation for each view on business rules.

Figure 1-1 gives an overview on the *BROCOM* approach. It shows the distinction of three main parts: the meta model, the modeling steps, and the repository.

In the research project these three elements have been put into a larger context which resulted in the following research topics:

- The *BROCOM* approach sketched above has to be regarded in a larger context; thus, in Chapter 2 closely related topics are discussed: the *organizational theory* and *implementation alternatives*. The organizational theory is analyzed with a focus on the formalization of business rules. The discussion of implementation alternatives considers different generations of programming languages and rules in DBMS.

---

[31] Cf. Nazareth (1993), p. 403.

[32] This is also supported by Navathe (1992), p. 5.

[33] In Balzert (1993), p. 52, the use of rules as basic technique is assigned to the analysis as well as to the implementation task of IS development.

[34] Cf. Bubenko/Wangler (1992), pp. 398; Habermann/Leymann (1993), pp. 19; Romberg (1995), p. 58; an application of a repository in the field of rule-based expert systems is given in Jansen/Compton (1989); for a comprehensive discussion of concepts and practical use of repository systems cf. Myrach (1995).

- While the second chapter covers the wide context, Chapter 3 focuses on business rules themselves. First, *case studies* are presented which are based on practical IS and reveal detailed information on the relevance of business rules. Second, a classification system is presented which should support the administration of business rules in *BROCOM*. Third, business rules are related to relevant organizational facts, e.g., processes, processors, or organizational units. Finally, a syntax for the formalization of business rules is introduced.
- After having provided the basis for *BROCOM*, the approach is introduced in Chapter 4. The presentation of *BROCOM* is done in two main parts: the first encompasses the description of the meta model and the second the specification of modeling steps.
- One of the assumptions listed above was that no new graphical models ought to be developed. In Chapter 5, *selected graphical models* for views on the meta data are presented and evaluated with respect to their applicability.
- In Chapter 6, the implementation of *BURRO*, a business rule repository system is described which is designed to support *BROCOM*. In order to illustrate the practical applicability of the approach, the *BURRO* system is applied to a large case study.

Finally, the work is concluded in Chapter 7 with a summary and outlook. In the appendixes, the syntax for business rules and details on the implementation of the *BURRO* system can be found.

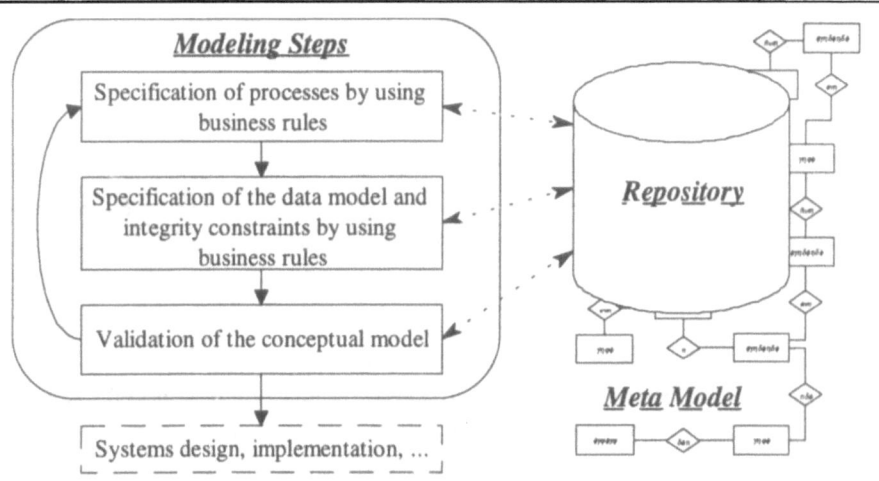

Figure 1-1: Framework of BROCOM: a business rule-oriented conceptual modeling

2 RULES IN ORGANIZATIONS AND INFORMATION SYSTEMS

Before focusing on business rules and presenting the *BROCOM* approach, two closely related topics will be regarded: business rules in the organizational theory and alternatives for the implementation of business rules in IS.

- The *organizational theory* is discussed to provide the context for the conceptual modeling of business rules. Because this modeling may result in the formalization of business rules which may have been non-formalized, this aspect is emphasized. The discussion of the formalization is first done with respect to four types of organizational approaches and afterwards from a general approach independent point of view.
- The *implementation of IS* is discussed with a focus on implementation alternatives which encompass different generations of programming languages and rules provided by deductive and active DBMS.

Both aspects of rules, the organization and the implementation, provide a basis for the approach for a conceptual modeling of business rules which is subject of the following chapters.

2.1 Rules in Organizational Theory

2.1.1 Introduction

The design of organizations is a problem concerning researchers as well as practitioners. Though the term *organization* is well known and often used, its definition is not standardized. Based on an analysis by Thompson[1], Galbraith states that

„first, organization emerges whenever there is a shared set of beliefs about a state of affairs to be achieved and that state of affairs requires the efforts of more than a few people. That is, the relationships among the people involved become patterned. The behavior patterns or structure derive from a divison of

---

[1] Thompson (1967), p. 52.

labor among the people and a need to coordinate the divided work. Thus, a primary contribution of organization structure is to coordinate the interdependent subtasks which result from the division of labor. Second, the analysis and the points just made above introduce the essential attributes of organization and allow us to define what we mean by the term organization. We can say that organizations are (1) composed by people and groups of people (2) in order to achieve some shared purpose (3) through a divison of labor (4) integrated by information-based decision processes (5) continuously through time."[2]

This view of organizations assumes that they emerge naturally in order to achieve goals, e.g., synthetic organizations which arise after a disaster[3]. This leads to the question: why design and formalize organizations? The main reason lies in the efficiency of the organization. Galbraith mentions that „if the synthetic organization is evaluated on its performance, it would be rated effective but inefficient"[4]. Therefore, the structure of organizations has to be analyzed and designed; this involves the formalization of certain facts. The need for formalization results from specific properties of organizations or as Etzioni states:

„The artificial quality of organizations, their high concern with performance, their tendency to be far more complex than natural units, all make informal control inadequate and reliance on identification with the job impossible. Most organizations most of the time cannot rely on most of their participants to internalize their obligations, to carry out their assignments voluntarily, without incentives"[5].

This argumentation is based on rather negative assumptions on the human nature; however, it covers the worst case. Therefore, assuming there are well motivated organization participants, a lesser, but certainly not an absence of formalization is appropriate.

The aspect of formalization is even part of a definition of the term *organization* given by Kieser and Kubicek, who regard organizations as social units which (1) lastingly want to achieve common goals and (2) have a *formal structure* which coordinates the activities of all members in order to achieve the goals.[6] The term *formalization* is used in the organizational theory for the written determination of

---

2   Galbraith (1977), p. 3.
3   Cf. Galbraith (1977), p. 2.
4   Galbraith (1977), p. 4.
5   Etzioni (1963), Chapter 6.
6   Cf. Kieser/Kubicek (1992), p. 4.

organizational facts by using e.g. business rules[7]. Within the following subsections, selected organizational approaches are introduced and evaluated with respect to formalization. Afterwards, the aspect of formalization is discussed independently from specific approaches[8].

2.1.2 Selected Organizational Approaches

In history, several organizational approaches have emerged which differ in emphasizing specific aspects of organization. The discussion of selected organizational approaches is done according to a classification presented by Kieser and Kubicek (cf. Figure 2-1). For each of the classes at least one representative is discussed with respect to the relevance of formalized rules.

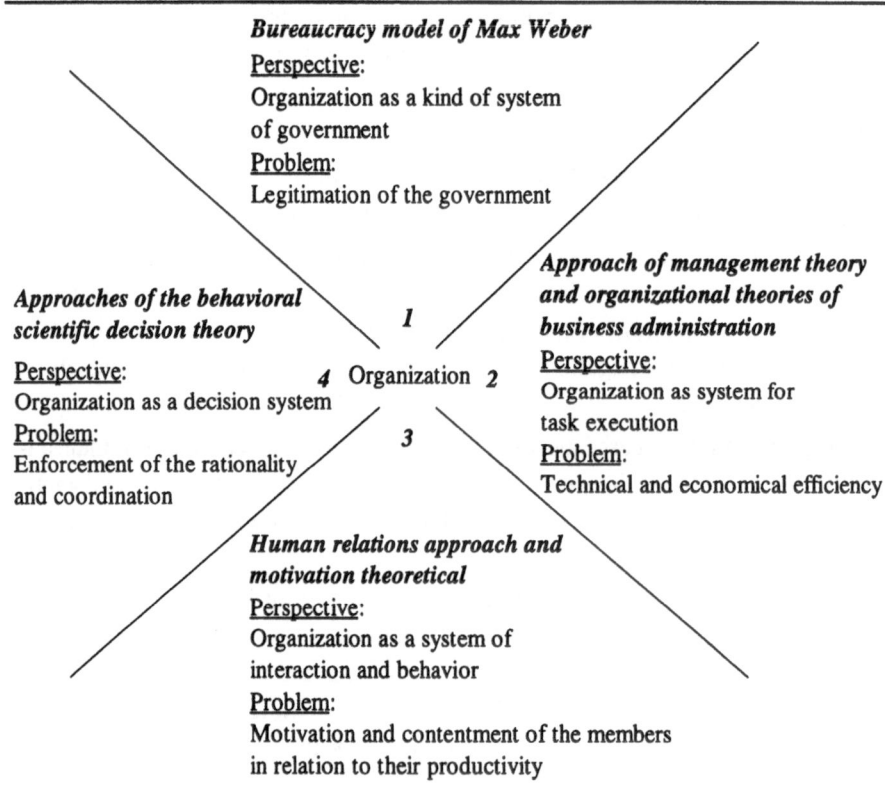

Figure 2-1:   Classification of organizational approaches[9]

---

[7]   Cf. Schmidt (1985), p. 38.

[8]   Additional aspects of rules, as e.g., their efficiency are discussed in Suter (1995).

[9]   Cf. Kieser/Kubicek (1992), p. 34.

2.1.2.1 Bureaucracy Model of Max Weber

The term *bureaucracy* has been introduced by Weber as an ideal type of structure and Weber intended the term as a purely technical one, as it is still considered in literature of the organizational theory[10]. However, elsewhere the term has got a pejorative meaning because the approach is closely linked to modern bureaucracies and insinuates 'civil service mentality'. Perrow states that

„'Bureaucracy' is a dirty word (...) It suggests rigid rules and regulations (...), impersonality, resistance to change. Yet every organization of any significant size is bureaucratized to some degree or, to put it differently, exhibits more or less stable patterns of behavior based upon a structure or roles and specialized tasks"[11]

The main characteristics of bureaucracy in its pure technical sense have been put together by Gerth and Mills:

„I. The principle of fixed and official jurisdictional areas, which are generally ordered by rules, that is, by laws or administrative regulations. (...)

II. The principles of office hierarchy and of levels of graded authority mean a firmly ordered system of super- and subordinate in which there is a supervision of the lower offices by the higher ones.

III. The management of the modern office is based upon written documents, which are preserved in their original or draught form.

IV. Office management, at least all specialized office management, (...) usually presupposes thorough and expert training.

V. The management of the office follows general rules, which are more or less stable, more or less exhaustive, and which can be learned. Knowledge of these rules represents a special technical learning which the officials possess. It involves jurisprudence, or administrative or business management." [12]

In their Aston studies, Pugh et al. [13] analyzed 52 organizations in the English Midlands in order to derive a classification and a taxonomy of structures of work organizations. Originally, 64 scales were used to operationalize the five primary variables:

---

[10] Cf. Weber (1972), pp. 551.

[11] Perrow (1970), p. 50.

[12] Gerth/Mills (1958), pp. 196.

[13] Cf. Pugh et al. (1969), pp. 115.

1. specialization of functions,
2. standardization of procedures (the existence of rules purporting to cover all circumstances and applying invariably),
3. *formalization of documentation* (the extent to which rules, procedures, instructions, and communications are written),
4. centralization of authority, and
5. configuration of positions.

Based on these variables, three dimensions of a taxonomy were derived, which resulted in the identification of seven types of bureaucracies (cf. Figure 2-2). Pugh et al. hypothesize about developmental sequences of the two dimensions: *structuring of authority* and *line control of workflow*. The first of these two dimensions seems to be dependent on the size of an organization. A correlation of 0,67 between size and structuring „strongly suggests that increasing structuring is concomitant with increasing size"[14]. From a long-run trend of increasing mechanization and standardization of products, the impersonality increases as control passes from the individual production worker and his direct supervisors to procedures dictated by standardization. The development of bureaucracies is described by Pugh et al. as follows:

> „With the nascent workflow bureaucracy there are the beginnings of structuring, the appearance of specialists and the expansion of procedural control. With the workflow bureaucracy, specialists appear, producing more procedures and reinforcement control by the line with impersonal bureaucratic regulations. The same sequence of arguments would apply in regard to the development of nascent full bureaucracies to full bureaucracies."[15]

Cluster	Structuring of activities	Line control of workflow	Concentration of authority	Number of organizations
Full bureaucracy	High	Low	High	1
Nascent full bureaucracy	Low	Low	High	4
Workflow bureaucracy	High	Low	Low	15
Nascent workflow bureaucracy	Mid	Low	Low	5
Preworkflow bureaucracy	Low	Low	Low	11
Personnel bureaucracy	Low	High	High	8
Implicitly structured organizations	Low	High	Low	8

Figure 2-2: Clusters of organizations[16]

---

[14] Pugh et al. (1969), p. 124.
[15] Pugh et al. (1969), p. 124.
[16] Pugh et al. (1969), p. 121.

The arrows in Figure 2-2 visualize the assumed developments of the two dimensions. As mentioned by Pugh et al., no development sequence applies to the third dimension *concentration of authority*.

Summarizing the discussion above, formalized rules are a very important component of Weber's bureaucracy approach which has strongly influenced today's organizations. However, there are also organizations which can be classified as *organic* and which work well especially in unstable environments which require the ability to innovate or to adapt to changing environments. Organic organizations which can be defined by „the absence of standardization on the organization"[17] therefore represent the other end of a continuum of formalization. The coordination in these organizations has to be provided by other mechanisms, e.g., those discussed later in the human relations approaches.

2.1.2.2 Approach of Management Theory and Organizational Theories of Business Administration

Management theory and business administration do not encompass a single but a rather large number of organizational approaches. Two well-known representatives are *scientific management* and *business process reengineering*. The first emerged in the beginning of industrialization and the second represents a current trend of organizational reflections.

2.1.2.2.1 'Scientific' Management

Beginning with industrialization, the principle or division of labor has become predominant which leads among others to the organizational approach of *'scientific management'*[18]. The basic assumption of scientific management is that the only objective of a worker is to maximize his income; thus, the consideration of sociological and psychological aspects may be neglected. The main goal of scientific management is to improve the efficiency and productivity of the workers by separating *reflection* and *execution*. Management has to find the optimal way of task execution and provide the worker with detailed instructions. Scientific management is based on four principles[19]:

- For each elementary task the old rules of thumb (traditional knowledge, intuition, and improvisation) are replaced by science (organizational rules, principles, and detailed procedures).
- Workers are selected and trained based on scientific studies.

---

[17] Mintzberg (1979), p. 87.
[18] Cf. Taylor (1911).
[19] Cf. Taylor (1911); Pohl (1981), p. 9.

- A close cooperation of managers and workers ensures that all tasks are performed with respect to the scientific instructions.
- The responsibility and the work to be done is to be distributed equally between management and the worker.

The application of these four principles led e.g. to movement and time studies which resulted in the optimal way of execution and the optimal duration of each elementary task. Based on this, the pay per unit was determined. Because of its orientation towards a very detailed task structuring, the application of scientific management leads to very detailed business rules prescribing every task to be performed. However, there exist criteria for an optimal level of structuring which are not considered in scientific management. Although, the approach resulted in major achievements for today's organizations, its basic assumption to neglect sociological and psychological aspects[20] and its goal to determine salaries from optimal task execution was criticized by e.g. the representatives of sociologically oriented organizational approaches.

2.1.2.2.2   Business Process Reengineering

Information technology (IT) has strongly influenced the organizations in the second half of this century. In *'The corporation of the 1990s'* Venkatraman discusses five levels of business reconfigurations resulting from the use of IT[21] (cf. Figure 2-3). One of them is *business process redesign* (BPR) which, after the publication of *'Reengineering the Corporation'* by Hammer and Champy[22], has become a major buzzword of organizational reflections. The trigger of the new orientation is the change of the environment in which business is done:

„The reality that organizations have to confront, however, is that the old ways of doing business - the division of labor around which companies have been organized since Adam Smith first articulated the principle - simply don't work anymore. (...) Adam Smith's world and its way of doing business are yesterday's paradigm."[23]

---

[20]   Cf. Gutenberg (1962), p. 145.

[21]   Venkatraman (1991), pp. 126.

[22]   Cf. Hammer/Champy (1993).

[23]   Hammer/Champy (1993), p. 17.

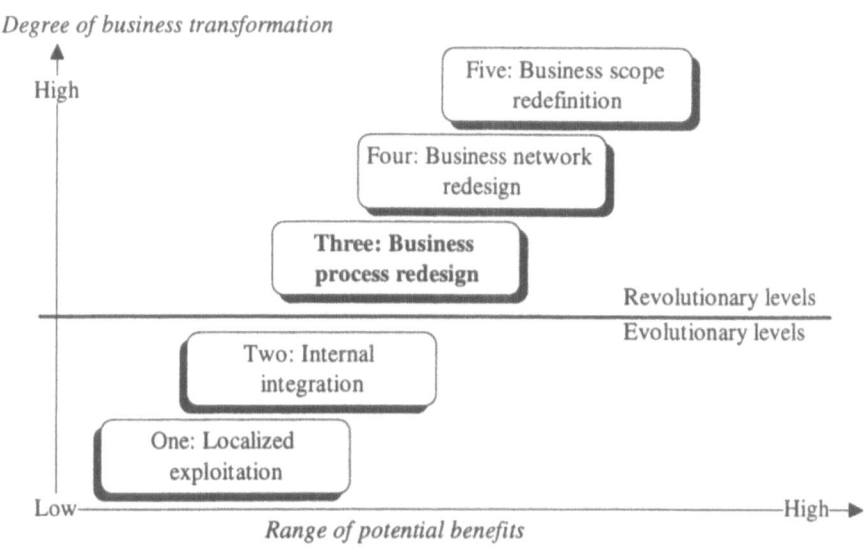

Figure 2-3:   Five levels of IT-induced reconfiguration[24]

Three major forces govern today's competition: *customers* who take charge, *competition* that is intensifying, and *change* that has become constant[25]. Thus, *market pressure* is seen as an enabler of a business process orientation[26]. The core message of Hammer and Champy is that „task-oriented jobs in today's world of customers, competition, and change are obsolete. Instead, companies must organize work around *process.*"[27]

The central term of the new organizational orientation is *reengineering* which is defined as „the fundamental rethinking and radical redesign of business processes to achieve dramatic improvements in critical, contemporary measures of performance, such as cost, quality, service, and speed."[28] The term process is defined as „a collection of activities that takes one or more kinds of input and creates output that is of value to the customer"[29]. In the context of *process innovation* a process is „a specific ordering of work activities across time and place, with a beginning, an end, and clearly identified inputs and outputs: a structure for ac-

---

[24]   Venkatraman (1991), p. 127.
[25]   Cf. Hammer/Champy (1993), pp. 17.
[26]   Cf. Venkatraman (1991), p. 140.
[27]   Hammer/Champy (1993), pp. 28.
[28]   Hammer/Champy (1993), p. 32.
[29]   Hammer/Champy (1993), p. 35.

tion"[30]. Both definitions imply that a business process collapses the value-added chain[31] (cf. Figure 2-4) and 'cuts' through traditional organizational structures (cf. Figure 2-5).

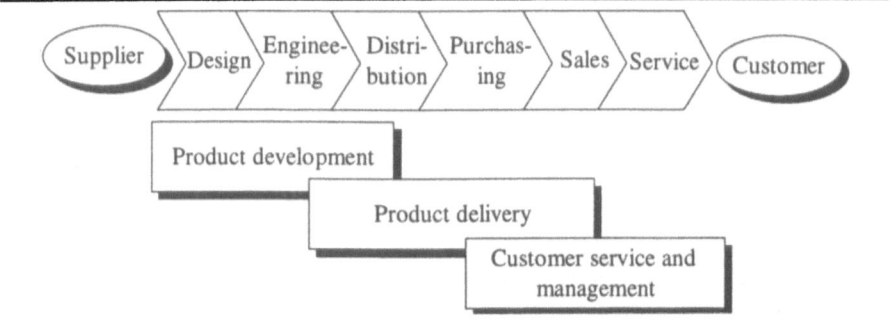

Figure 2-4:   Product development, product delivery, and customer service and management: Collapsing the value-added chain[32]

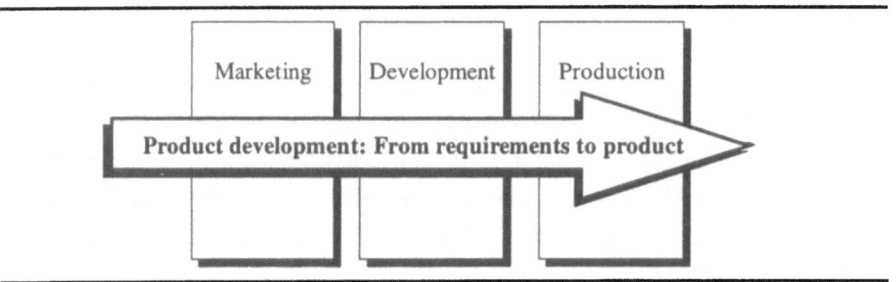

Figure 2-5:   Processes cut straight through the functional organization structure[33]

One main advantage of process orientation is the measurability of processes which have cost, time, output quality, and customer satisfaction[34] and can therefore be designed and evaluated.

In the context of this chapter, the formalization of organizational facts within BPR is of interest; however, this is a topic widely neglected in literature on BPR. In the following, the formalization aspect of the approach is discussed based on goals and models. First, arguments for a lower formalization are presented, followed by arguments for a higher one.

---

[30] Davenport (1993), p. 5.
[31] Cf. Rockart/Short (1991), pp. 207.
[32] Rockart/Short (1991), p. 207.
[33] Jacobson et al. (1995), p. 4.

Lower Formalization

One main goal of BPR is to achieve more *flexibility* by e.g. reducing checks and controls, delegating decision making, and providing several versions of a process[35]. This need for organizational flexibility is certainly not supported by a high degree of formalization.

A lower formalization also results from two properties of business processes: their *high complexity* and their *instability* over time. Regarding methodologies of IS development, it can be seen that the modeling of data structures is well solved by standardized approaches like the Entity-Relationship Model[36], whereas the modeling of processes is still not supported by any widely accepted and standardized approach. One reason is that data structures are generally less complex and more stable in time and can therefore be formalized more easily. A detailed specification of business processes, however, has to consider very complex and unstructured facts which additionally have to be adapted often, in order to keep the model consistent with the real world. Therefore, a rather general and 'imprecise' modeling of business processes, i.e., a lower level of formalization, would be appropriate.

Higher Formalization

Important elements of most BPR approaches are the claimed possibilities to evaluate business processes by e.g. measuring and simulating them. In this context, Denna et al. state that you „can't change what you can't measure"[37]; however, in order to not only estimate, but to really *measure* a business process with respect to e.g. time and cost, the process has to be specified in detail. According to Morris and Brandon „these evaluations are made to detailed function and process designs" [38]; thus, they can be very accurate. They further say „that the work carries the (...) models and data to additional levels of detail and refines the data so that the problem areas and interrelationships are visible in full detail"[39]. Thus, they regard formalization of business processes as a necessary basis for successful BPR. Davenport states that the „structural element of processes is key to achieving the benefits of process innovation. Unless designers or participants can agree on the way work is or should be structured, it will be very difficult to

---

[34] Cf. Davenport (1993), p. 6; Hammer/Champy (1993), pp. 72; Jakubczik/Skubch (1994), pp. 61.

[35] Cf. Hammer/Champy (1993), pp. 53.

[36] Cf. Batini et al. (1992).

[37] Denna et al. (1995), p. 353.

[38] Morris/Brandon (1993), p. 175.

[39] Morris/Brandon (1993), p. 171.

systematically improve, or effect innovation in that work"[40]. He suggests the specification of detailed *process* flows on three levels[41]:

- *Process* level (inputs, outputs, interfaces, flows, and measures).
- *Subprocess* level (objective, performance metrics, who performs, IT enabler, information needs/activities, value-added, and activities in the process).
- *Activity* level (information needed, decision point, who does it, value-added).

Another aspect is the validation of business processes which may be supported by *simulations*[42]. A simulation of processes may be based on rather general process specifications. However, such a simulation does not tackle the core problem, because it can be assumed that a general and not very detailed specification may be one on which agreement can be achieved easily. A detailed specification which encompasses more complex activities and interdependencies is harder to understand and to agree on and is therefore the appropriate level for the use of simulations. This again demands for a detailed process formalization[43]. In literature several business models have been proposed[44] and Carr and Johansson put together a list of data which should be included in such a model[45]:

- Process flow data (activities and tasks).
- Resource data (people, machines, support systems, and business units).
- Input data (product mix, demands, and volumes).
- Event data (when events occur, e.g., meetings or accounts closing).

Business processes are not only evaluated by measuring and simulating them, they also have to be further analyzed before being definitely implemented in an organization. These analyses should among other things detect redundancies in activities and processes, bottlenecks in work and task flows, inefficiencies, or inappropriate or unclear interfaces[46]. In order to achieve this, again the process specifications have to be very detailed.

In Davenport's discussion of the role of information technologies (IT) within BPR, he points out *automation* as one major opportunity resulting from the use of IT:

---

[40] Davenport (1993), p. 5.
[41] Davenport (1993), p. 139.
[42] Cf. Carr/Johansson (1995), pp. 150.
[43] Cf. e.g. Ferscha (1994), pp. 223.
[44] Cf. e.g. Morris/Brandon (1993); Jacobson et al. (1995).
[45] Carr/Johansson (1995), p. 150.
[46] Cf. Morris/Brandon (1993), p. 176.

„This opportunity, long understood in manufacturing, is the province of robotics, cell controllers and so forth. In service environments, where processes are frequently defined by document flows, automational opportunities increasingly rely on imaging systems that remove paper from the process, frequently accompanied by 'work flow' software that defines the path images follow through a process."[47]

If (partial) automation of business processes is regarded as a goal of a BPR project, of course a higher formalization of processes becomes absolutely necessary.

Consolidating the pros and cons of formalization in BPR, it can be said that on the one hand, a low formalization is to be achieved in order to have a flexible organization. On the other hand, business processes should be formalized in detail in order to be able to evaluate, measure, analyze, and simulate them. One solution of this goal conflict might be that business processes should be described in detail during their design and that the implementation of business processes in the organization could make use of a reduced formalization. However, this could lead e.g. to totally different times and costs, since the process executed in reality is not equivalent to the one used in the evaluation. Because of this gap, the evaluation of process alternatives becomes pointless and, since this evaluation is a major task of a BPR project, the whole project may become superfluous.

Higher formalization	Lower formalization
Measurability of process design	Flexibility of process execution
Analyzability of process design	Complex reality
Comparability of process design	Motivation of people
Automatability of process execution	

Figure 2-6: Pros and cons of a higher formalization in BPR

As will be shown in Chapters 3 to 6, business rules may be very useful for the specification of certain aspects of business processes and workflows, e.g., details of activities and their relationships with data, organizational units, and manual or automated processors[48].

---

[47] Davenport (1993), p. 51.

[48] The use of business rules in the context of business process reengineering is also discussed e.g. in Morris/Brandon (1993), pp. 129; Appleton (1995), pp. 291.

2.1.2.3   Human Relations Approach and Theory of Motivation

The origin of the *human relations approaches* are the experiments of Pennock which took place in the Hawthorne site of the Western Electric Company. These experiments investigated the influence of the quality of light on the productivity of workers[49]. The results first indicated a dependency of productivity on the working conditions; however, it had been recognized that the improvements did not only result from better working conditions, but also from the fact, that the people knew that they were part of such an experiment and that they were observed. These findings triggered further research in the area of the relationship between contentment and motivation of people and their productivity[50]. The basic model of the human relations approach is depicted in Figure 2-7. Regarding rather formal approaches as bureaucracy and scientific management, the human relations approach recognized the need for unregulated and unplanned behavior in order to manage exceptions in unexpected situations. However, the early approaches totally ignored the problem of formalizing organizational facts[51]. Later, the coexistence of formalized structures and individual relations was investigated[52].

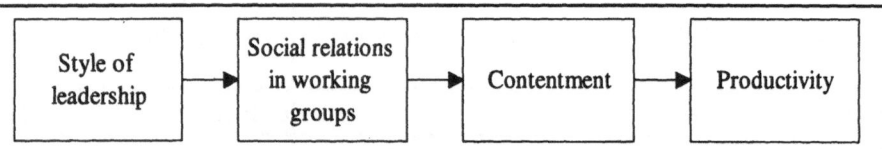

Figure 2-7:   Basic model of human relations approaches[53]

2.1.2.4   Approaches of the Behavioral Scientific Decision Theory

The approaches of the behavioral scientific decision theory regard organizations as cooperative systems in which individual decisions have to be coordinated in order to achieve the organization's goals. This coordination and thus the existence of an organization depends on the (at least partial) acceptance of its goals by the individuals. Each individual personality has to be dominated by the organization's personality and it is the primary function of managers to establish

---

[49] Cf. Roethlisberger/Dickson (1939).
[50] Cf. Kieser/Kubicek (1992), p. 40.
[51] Cf. Pohl (1981), p. 25.
[52] Cf. Argyris (1957); McGregor (1960); Argyris (1962); Bennis (1966); Likert (1967); Bennis (1969).
[53] Cf. Pohl (1981), p. 24.

these organizational personalities within the individuals. The work that an individual does for an organization is seen as a decision problem in which the individual evaluates different alternatives from which hopefully the ones are chosen which support the organization's goals. The decisions of the individuals are very complex especially because of incomplete knowledge on existing decision alternatives and criteria. The organization can influence these decisions by *providing incentives* or by *influencing the perception and evaluation* of the incentives by an individual. The incentives proposed are among others monetary incentives, personal non-monetary incentives (like prestige or power), the physical design of the working-place, and some indirect incentives which are often called working atmosphere[54]. In order to influence the perception of incentives, the organization may use formalization, rationalization of opportunities, and teaching specific values in an early process of socialization. The individual's decisions not only have to be influenced but also must be coordinated. This can be achieved by one of three basic alternatives[55]:

- Members of an organization coordinate their activities in negotiations.
- Superiors coordinate the work of their subordinates by appropriate instructions.
- Coordination of decisions by the application of rules which have been coordinated and made consistent.

Thus, the approaches which are assigned to this class of organizational approaches do not emphasize formalization but may make use of it in order to coordinate the decision-making process of the individuals. However, the focus is on the decision coordination and therefore the question of formalization is not crucial within these approaches.

2.1.2.5 Comparison of the Selected Organizational Approaches

The organizational approaches previously discussed can be positioned on an axis ranging from a high level of formalization (scientific management and bureaucracy model) to a lower one (human relations approaches). Consistently, Bennis characterizes scientific management and bureaucracy as *organizations without people* and the approaches of human relations as *people without organizations*[56]. According to the discussion, approaches of the behavioral scientific decision theory are considered as rather informal and the concepts of BPR as rather formal.

---

[54] Cf. Kieser/Kubicek (1978), p. 43.
[55] Cf. Kieser/Kubicek (1978), p. 50; Kieser/Kubicek (1992), p. 42.

Summarizing the discussion of the selected approaches, it can be said that the problem of finding the optimal level of formalization is still an open question of organizational reflections and that the extreme positions certainly do not provide an appropriate and realistic solution. In the following section, a more general, approach-independent view on formalization is attempted.

2.1.3   Formalization of Organizational Facts

Organizations are usually complex structures which become manageable if they are controlled by more or less formalized rules. The oldest known formalization of organizational rules was found in old Egypt and describes tasks needed for the construction of the pyramids[57]. Another old and very large set of formalized rules was specified for the administration of the Chinese Empire of the Chou Dynasty. These guidelines were published in 1100 B.C. and included rules for all departments of the empire[58].

In our cultural area, non-written rules were dominant until the late 60's of the 19th century. Since the end of the 19th century, the division of labor has become more important resulting in a more complex net of interactions in business processes. To coordinate elementary tasks and to achieve a more constant execution of theses tasks, formalized rules have begun to replace non-written ones.

Today, the documentation of formalized business rules is often done using an organizational handbook which contains systematically collected and specified business rules and is often considered as the 'law of the organization'[59]. Organizational handbooks may be structured as follows[60]:

1. *Overview*: Motivation for the formalization of rules in the organization.
2. *Organizational structures* are described by e.g. graphical schemes (including positions within the organization), verbal descriptions of duties and rights of employees, or rules on authorizations to sign.
3. *Processes* may be divided in subprocesses and described in e.g. a graphical model including e.g. task dependencies, time restrictions, execution responsibilities, and exception handling.
4. *Appendixes* include e.g. the standardization of terms, forms, and abbreviations used in the organization.

---

[56] Cf. Bennis (1959).
[57] Cf. George (1972), pp. 4.
[58] Cf. George (1972), p. 12.
[59] Schmidt (1989), pp. 336.
[60] Cf. e.g. Schmidt (1989), p. 337.

In the following, the discussion focuses on business rules encompassing knowledge on organizational dynamics; rules determining the organizational structure are not further considered.

The standardization and formalization of processes serves to coordinate operations of an organization[61]. Mintzberg distinguishes three types of behavior formalization[62]:

- *Formalization by job*: In this case, the formalization is attached to a specific job and is normally included in the formal job descriptions. Examples are given by March and Simon[63]:

    1. *When material is drawn from stock, note whether the quantity that remains equals or exceeds the buffer stock. If not:*

    2. *Determine from the sales forecast provided by the sales department the sales expected in the next k months.*

    3. *Insert the quantity in the 'order quantity formula' and write a purchase order for the quantity determined.*

- *Formalization by work flow*: Instead of linking them to the job, the organization can attach the formalization to the work itself.
- *Formalization by rules*: Formalization by rules is a more general approach which may institute rules for all situations, i.e., all jobs and work flows. These rules may „specify who can or cannot do what, when, where, to whom, and with whose permission"[64].

The management of an organization with formalized rules restricts the space to be oneself and the responsibilities of employees. This contradicts the social trend towards the demand to act on one's own responsibility. Morgan states that „we are leaving the age of organized organizations and moving into an era where the ability to understand, facilitate, and encourage processes of self-organization will become a key competence"[65]. Self-organization can be seen as extreme non-formal and, in general, is a main property of systems[66]. Formalized rules can be used either to accelerate or to slow down the process of self-organization in systems[67]. Therefore, self-organization is not an organizational approach on its own,

---

[61] Cf. Mintzberg (1979), p. 7.
[62] Cf. Mintzberg (1979), pp.81
[63] Cf. March/Simon (1958), pp. 147.
[64] Mintzberg (1979), p. 82.
[65] Morgan (1993), foreword.
[66] Cf. Probst (1987), p. 11.
[67] Cf. Probst (1987), p. 88.

but has always been a (sometimes unintended) part of organizations, resulting in informal information exchange or groups which constitute themselves and become active on their own. Reasons for this include among others quantitative, qualitative, and acceptance problems:

- *Quantitative problems*: It is impossible and not adequate to formally specify every fact and activity of an organization.
- *Qualitative problems*: Organizations are too complex to be fully specified in rules.
- *Acceptance problems*: Every rule has to be accepted by the employees because it restricts their activities. Therefore, rules may be regarded as not necessary and will either not be followed or can result in a lower motivation of the employees.

In addition to these problems, the use of formalized rules may trigger vicious circles which also are discussed in literature[68]. Figure 2-8 depicts the one of Merton as presented by March and Simon. In this circle, only the mutual influence between defensibility of individual actions and emphasis on reliability is intended. The other nodes and relationships of the circle are unintended and may lead to a dysfunction of the organization resulting from a rising level of formalization.

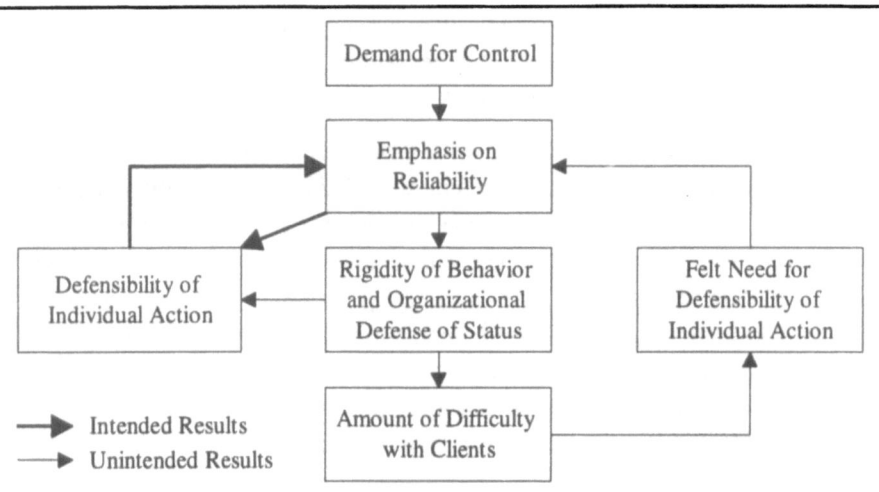

Figure 2-8:   *Consequences of excessive behavior formalization: „The simplified Merton Model"[69]*

---

[68] For further details on the dysfunction of highly formalized structures cf. e.g. Crozier (1964); Mintzberg (1979).

[69] Cf. March/Simon (1958), p. 41.

However, formalized rules are important elements of most existing and successful organizations in such a way that they simply have to be regarded as „decisions made in advance of their execution. That is, to the extent that decisions are repetitive, a procedure is worked out in advance of encountering the situation. The virtue of rules is that they eliminate the need for communication between interdependent parties and between superior and subordinate"[70]. The following list summarizes important reasons in favor of the use of formalized rules:[71]

- *Coordination*: In complex situations, the execution of tasks may need synchronization of the work done by several persons.
- *Precision*: The execution of as far as possible formalized tasks is precise over time, i.e., everybody knows what to do in every considered event.
- *Efficiency*: A machinelike consistency of the task execution may lead e.g. to efficient production processes as in the automobile industry.
- *Fairness*: Especially government organizations have to secure an equal treatment of every client; hence, they strive to formalize their behavior in order to protect clients and also employees.

Schmidt discusses the decision on the formalization with respect to *internal* and *external influences*. The most important *internal* ones result from the *properties of the task*[72]:

- *Repetition rate*: A task is executed only once or may be repeated several times. If a certain repetition can be assumed the task is suited for being formalized as business rules. The repetition rate normally increases with the size of an organization, whereas it often decreases towards the upper levels of the organization's hierarchy.
- *Constancy*: A specific task can be executed in different ways depending on a specific situation. In a very unstable and dynamic environment, task execution is mostly less constant and therefore the task should not be formalized. However, if the constancy of repeated tasks is high, its formalization may be appropriate.
- *Complexity*: The complexity can be measured in regard to the number of elementary tasks to be interrelated. The higher the complexity of an organization or a part of it is, the more need arises for coordination by e.g. formalization.

---

[70] Galbraith (1977), p. 43.
[71] Cf. Galbraith (1977), pp. 43; Mintzberg (1979), p. 83.
[72] Cf. Schmidt (1985), pp. 40.

- *Determinability*: The execution of tasks can be more or less determined in advance. One extreme are non-deterministic tasks, whereas others are fully determined and may even be automated.

Additionally, characteristics of the organization (e.g. industry, market position, legal form, age, size, or life cycle), the people (e.g. attitudes, capability, or motivation), and the technical environment (characteristics of production processes, transport facilities, computer support, or communication infrastructure) determine the use of formalized rules[73].

The *external* influences on the use of written rules result from actual or potential activities of direct or indirect interaction partners of the organization[74]. They may result from *competition, customer structure, market structure, technology*, and *social and cultural*.

The decision on an appropriate level of formalization strongly influences the development of an IS by restricting the use of fully automated processing to those parts of an organization which can and should be formalized in detail. However, especially in early stages of IS development, general facts which are not specified in details have also to be considered in order to get a comprehensive model of the universe of discourse.

2.2   Implementation of Rules in Information Systems

The preceding section dealt with organizational theory as a part of business administration. As mentioned in the introduction of this chapter, the second important aspect is the implementation of business rules which are relevant within an organization. Two main alternatives are considered:

- Program code written in third, fourth, or fifth generation languages
- Rules in deductive or active DBMS

The discussion focuses on how *business rules structured as ECA^2 rules* can be implemented using the constructs of the implementation alternative.

---

[73]   Cf. Grochla (1978), pp. 19.

[74]   Cf. Grochla (1978), pp. 18.

2.2.1   Rules in Program Code

The implementation of IS can be done using different types of programming languages which are classified into five generations[75]. In the following, first generation languages (machine code) and second generation languages (assembler) are not regarded as considerable alternatives for the implementation of business IS. The main characteristics of third, fourth, and fifth generation languages are put together in Figure 2-9[76].

| | Language Generation | | |
	Third / 3GL	Fourth / 4GL	Fifth / 5GL
Basic structures	Procedural	Non-procedural, declarative, set-oriented, English-like syntax	Rule and frame based
Programmer	Specialists	End user	Specialists (knowledge engineers)
Examples	ALGOL, BASIC, C, COBOL, FORTRAN, Pascal, ...	FOCUS, MANTIS, NATURAL, RAMIS, Powerbuilder, Uniface, Oracle*Forms, ...	OPS5, LISP, ...
Data stores	Flat files, databases	Databases	Knowledge bases

Figure 2-9:   Programming languages of the 3rd, 4th, and 5th generation

2.2.1.1   Third Generation Languages

Third generation languages are used to implement procedures consisting of *sequences* of program instructions and the two basic structural constructs *selection* (IF-THEN-ELSE) and *iteration* (REPEAT-UNTIL or DO-WHILE).

Using these constructs, the programmer has two alternatives for the implementation of business rules structured as ECA^2 rules:

1. Each business rule is implemented as a module.
2. Each (sub)process consisting of a set of business rules is implemented as a module.

These two alternatives are discussed with respect to the rule sequence depicted in Figure 2-10. A disadvantage common to both alternatives is the lack of an event construct. Therefore, dynamic links between the business rules have to be implemented by explicit and therefore static procedure calls which may lead to maintenance problems.

---

[75]   Cf. Disterer (1988), pp. 2.

[76]   The comparison is based on Disterer (1988); Bordoloi/Jenkins (1991).

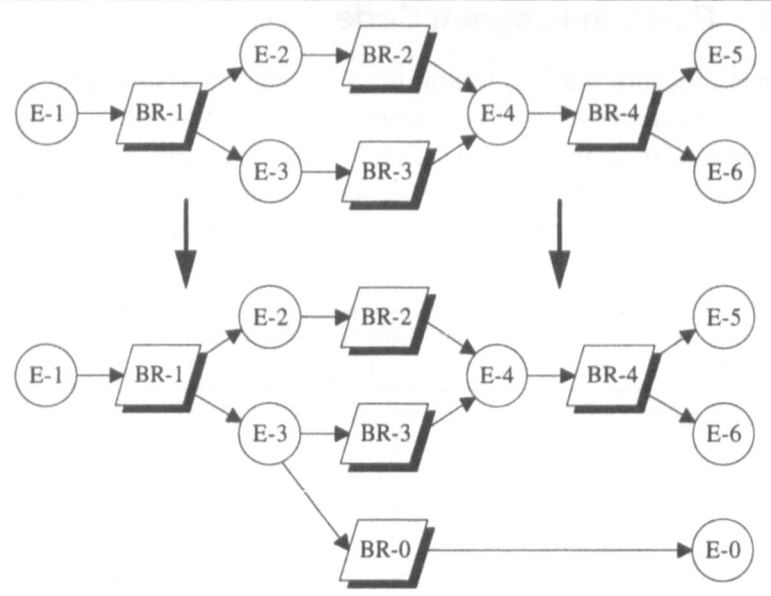

Figure 2-10: Example of rule sequences

Alternative 1

The modularization is based on single business rules and results in four modules, one for each of the business rules:

Rule-module 1: /* explicitly called by rule-modules which raise event E-1   */
IF (condition-1) THEN
    then-action-1
    call rule-module-2   /* replaces event E-2   */
ELSE
    else-action-1
    call rule-module-3   /* replaces event E-3   */
END IF

Rule-module 2: /* explicitly called by rule-module-1   */
IF (condition-2) THEN
    then-action-2
    call rule-module-4   /* replaces event E-4   */
 END IF

Rule-module-3: /* explicitly called by rule-module-1   */
IF (condition-3) THEN
    then-action-3
    call rule-module-4   /* replaces event E-4   */
END IF

Rule-module-4: /* explicitly called by rule-module-2 and rule-module-3   */
IF (condition-4) THEN
    then-action-4
    call rule-modules triggered by event E-5   /* not part of the example   */
ELSE
    else-action-4
    call rule-modules triggered by event E-6   /* not part of the example   */
END IF

The modification of the process by inserting *BR-0* would lead to the modification of *rule-module-1* and the creation of a new module:

Rule-module-1: /* explicitly called by rule-modules which raise event E-1   */
IF (condition-1) THEN
    then-action-1
    call rule-module-2   /* replaces event E-2   */
ELSE
    else-action-1
    call rule-module-3   /* replaces event E-3   */
    call rule-module-0   /* replaces event E-3   */
END IF

Rule-module-0: /* explicitly called by rule-module-1   */
IF (condition-0) THEN
    then-action-0
    call rule-modules triggered by event E-0   /* not part of the example   */
END IF

The advantage of this implementation alternative is a modularization which is congruent with the rule specification on the conceptual level. Therefore, the modules can be combined very flexibly. A disadvantage is the large amount of modules to administer.

Alternative 2

The implementation of a whole (sub)process as a module leads to a sequential-ized execution of business rules within a procedure. As an example the following program structure for a module specifying the (sub)process of Figure 2-10 is given:

```
IF (condition-1) THEN                        /* business rule BR-1        */
    then-action-1
    IF (condition-2) THEN                    /* business rule BR-2        */
        then-action-2
    END IF
ELSE
    else-action-1
    IF (condition-3) THEN                    /* business rule BR-3        */
        then-action-3
    END IF
END IF

IF (condition-4) THEN                        /* business rule BR-4        */
    then-action-4
    call modules which are triggered by E-5  /* not part of the example   */
ELSE
    else-action-4
    call modules which are triggered by E-6  /* not part of the example   */
END IF
```

Again, the insertion of *BR-0* into the subprocess is regarded. The main problems in incorporating this into the procedure are (1) the parallel execution of *BR-3* and *BR-0* and (2) that this parallelization leads to two different events at the end of the procedure, i.e., the module has one entry and two non-disjunctive exits. The parallel execution within a single procedure is not possible; therefore, *BR-3* and *BR-0* have to be artificially sequentialized. Of course the sequence of the two rules *BR-3* and *BR-0* is ambiguous, i.e., also the sequence *BR-0/BR-3* would be possible. Another disadvantage is the redundant implementation of a business rule if it is part of more than one process.

```
    IF (condition-1) THEN                    /* business rule BR-1        */
        then-action-1
        IF (condition-2) THEN                /* business rule BR-2        */
            then-action-2
        END IF
    ELSE
        else-action-1
        IF (condition-3) THEN                /* business rule BR-3        */
            then-action-3
        END IF
        IF (condition-0) THEN                /* business rule BR-0        */
            then-action-0
            call modules triggered by E-0    /* not part of the example   */
        END IF
    END IF


    IF (condition-4) THEN                    /* business rule BR-4        */
        then-action-4
        call modules which are triggered by E-5   /* not part of the example   */
    ELSE
        else-action-4
        call modules which are triggered by E-6   /* not part of the example   */
    END IF
```

Both alternatives for the implementation of business rules result in major problems and neither of them provides a flexibility which is equal to the event construct of business rules.

2.2.1.2   Fourth Generation Languages

Fourth generation languages (4GL) cannot unambiguously be distinguished from 3GL[77]. The term 4GL was used as a buzzword for the sale of programming languages and a rather large amount of programming languages which are claimed to be a 4GL are actually a 3GL enhanced by some declarative constructs like SQL statements. For the implementation of business rules, the use of a procedural 4GL leads to similar problems as the application of a 3GL; therefore, the discussion is restricted to declarative 4GL.

---

[77] In Pieper (1992) the distinction is that 3GL are procedural and 4GL object-oriented, whereas 4GL's are also characterized by the fact that they are non-procedural, cf. Martin (1985); Martin/Leben (1986); Misra/Jalics (1988); Bordoloi/Jenkins (1991).

Older 4GL's encompass FOCUS, RAMIS, MANTIS, and NATURAL[78], and newer products are e.g. Oracle*Forms, Powerbuilder, and Uniface. Especially the modern 4GL have adopted the notion of events; however, the events available in these 4GL are either events on a (graphical) user interface (e.g. a mouse click in a specified area or the leaving of a screen field) or events raised by data modification. Events related to the user interface may be relevant in business rules but result mainly from the design and implementation of the user interface and are therefore not considered as implementations of business rules. Events related to data modifications on the other hand are often relevant in the context of business rules. However, there exist types of elementary events (e.g. time points) or composite events[79] which cannot be detected by 4GL's.

2.2.1.3   Fifth Generation Languages

Since the beginning of interactive computing, researchers and practitioners have been trying to make computers 'intelligent'. This vision has led to intense research in artificial intelligence (AI)[80]. Because of the rule context, this section is restricted to rule-based programming which is used to build expert systems which are a special form of AI systems. In the following, the term expert system is used for „a computer system that performs functions similar to those normally performed by a human expert"[81]. Expert systems perform several generic tasks which may concern

- data *interpretation* (probably with missing or unreliable data values),
- *fault finding* (i.e. diagnosis),
- *monitoring* of signals or signal patterns,
- *prediction* of the course in the future from a model of the past and present,
- *planning* activities in order to achieve specified goals, and
- *designing* objects with respect to the requirements specified.[82]

Because these scopes of expert systems differ significantly from the scope of process- or data-oriented business IS, the business rule approach is not to be applied on the conceptual design of expert systems[83]. However, to clarify the differ-

---

[78] Cf. Martin (1995), pp. 290.

[79] Cf. the discussion of event algebras in Section 2.2.2.3.2 and the event classification in Chapter 3.

[80] For an overview of the developments in AI cf. e.g. Hayes-Roth (1988), p. 6.

[81] Goodall (1985), p. 10.

[82] Cf. Stefik et al. (1983), pp. 82.

[83] A software engineering approach for rule-based expert systems is described in Jacob/Froscher (1990).

ence between business rules and rules in expert systems, the basic structure of expert systems is sketched and a small example of expert system rules is given.

An expert system consists of three main elements (cf. Figure 2-11)[84]:

- *Knowledge base:* The knowledge base contains rules and facts about a particular application or problem domain and is created by knowledge engineers.
- *Inference engine:* The inference engine is activated when a user consults the expert system in order to seek a specified goal. In case of insufficient or imprecise information, the inference engine may ask the user to supply additional information which is added to the working memory.
- *Working memory of conclusions:* The working memory contains all information, preliminary conclusions, and results which are used for backward and forward chaining.

A rule in the knowledge base is normally in the form 'IF condition THEN action' where a condition is often a Boolean expression, e.g., „Color = Green", and the action a value assignment or a statement on a fact.[85]

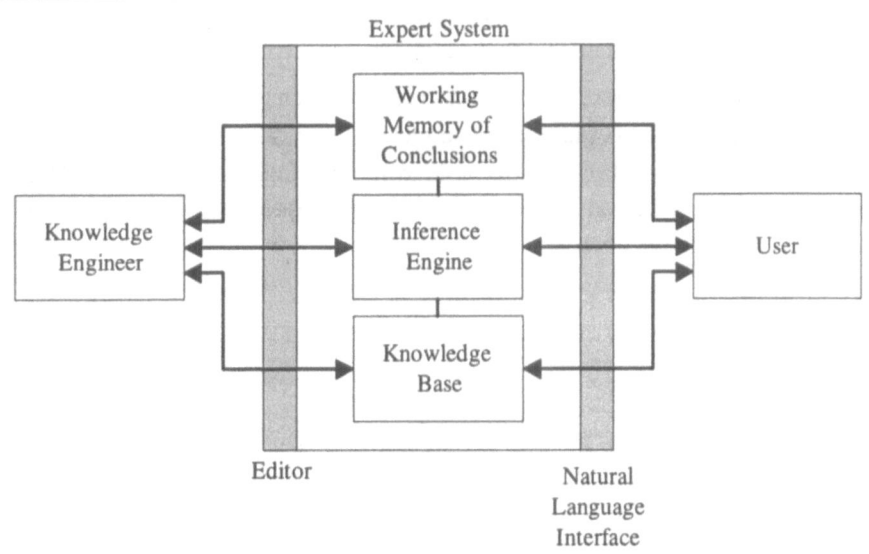

Figure 2-11: Illustrative expert system architecture[86]

---

[84] Silverman (1987), pp. 13.
[85] Silverman (1987), pp. 15.
[86] Silverman (1987), p. 14.

In order to illustrate rules in expert systems, the following simplified rules of
macro economics could constitute a knowledge base[87]:

IF   budget needs more money
THEN   tax increase

IF   the budget needs more money
THEN   real wages decrease

IF   tax increases OR real wages decrease
THEN   workers' incentives decrease

IF   workers' incentives decrease
THEN   production decreases

IF   production decreases
THEN   export decreases

IF   export decreases
THEN   budget needs more money

The inference engine could now be used to evaluate if, based on the known fact
'the budget needs more money', the assumption 'exports decrease' is true[88].

As already mentioned above, the business rule approach is not designed to be applied on expert systems because its basic idea is to describe the dynamic properties of an IS. However, because the basic syntax is similar, the structure of the
repository system described in Chapter 6 could be adapted in order to administer
but not to execute rules of an expert system.

---

[87] Cf. Goodall (1985), p. 43.
[88] For details on the reasoning for this example cf. Goodall (1985), pp. 43.

2.2.2 Rules in Database Management Systems

2.2.2.1 Overview on Current Research

The specification and implementation of DBMS as a research area dates back to the mid-60's[89]. In the current research in this field several main directions are followed[90]. On a symposium on the databases of the 90's, Lockeman et al.[91] considered four research directions as especially important. In the context of rules especially the second and third of them are of interest:

- Object-oriented DBMS,
- *deductive databases,*
- *active databases,* and
- parallel databases and database machines.

Consistent to this judgment, the Committee for Advanced DBMS Function[92] states in a proposition that „Rules (triggers, constraints) will become a major feature in future systems". Furthermore, „many participants pointed out the need for so-called active data bases. Hence, triggers, alerters, constraints, etc. should be supported by the DBMS as services. This capability was one of the needs most frequently mentioned as a requirement of future applications of data bases. There was extremely strong consensus that this was a fertile research area and that providing this service would be a major win in future systems"[93].

The research topic of incorporating rules in DBMS results from a combination of the research areas AI and DBMS (cf. Figure 2-12). The resulting rules are either active rules of *active DBMS* or deductive rules of *deductive DBMS*:

- *Active rules* specify the reaction of the system to the occurrence of certain situations. Active rules resulting from AI use pattern matching to trigger an active rule (production rules), whereas the database approach uses explicit events to activate database triggers.
- *Deductive rules* define a dataset and try to derive new data from existing facts. The AI approach for deductive rules are logic programs which are again based on pattern matching. Deductive rules from the database research are specified by views on existing data.

---

[89] Cf. Küng (1994a), pp. 58.
[90] Küng (1994a) gives an overview of current developments in the area of DBMS research and practice.
[91] Lockeman et al. (1990), pp. 22.
[92] Cf. Beech et al. (1991), pp. 498.
[93] Laguna Beach Participants (1989), p. 22.

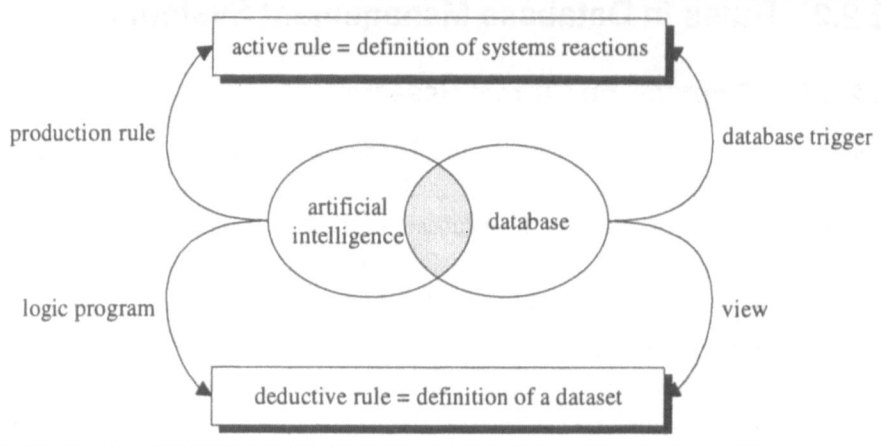

Figure 2-12: Basic research approaches for active and deductive rules[94]

Though this may be implied by Figure 2-12, the two approaches are not totally independent research directions but more extreme positions as depicted in an evaluation of six database rule languages[95] (cf. Figure 2-13).

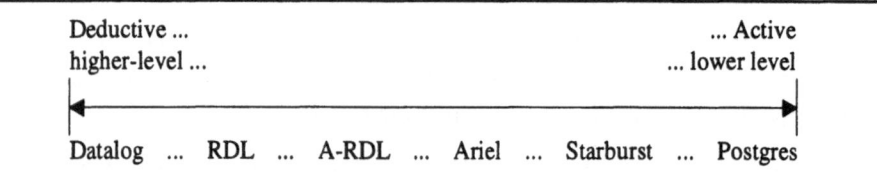

Figure 2-13: Spectrum of Database Rules Languages[96]

In the following an overview on deductive and active DBMS is given. This overview attempts to link the methodological discussions in the following chapters with rule aspects on the physical level.

---

[94] Cf. Manthey (1993).

[95] For a short overview on the rule languages cf. Widom (1993). For Datalog cf. Ullman (1989); for RDL cf. Kiernan et al. (1990); for A-RDL cf. Simon et al. (1992); for Ariel cf. Hanson (1989); for Starburst cf. Widom et al. (1991); for Postgres cf. Stonebraker et al. (1990).

[96] Widom (1993), p. 307.

2.2.2.2 Deductive Database Management Systems

Deductive DBMS have emerged from the relationship between the relational data model and logic programming[97]. The original problem was how to deduce facts from other facts to describe e.g. derived data. First work in this area already appeared in 1978 at the conference on Logic and Databases[98]. A definition widely used today defines deductive databases as „a database in which new facts may be derived from facts that were explicitly introduced."[99] From this definition the link between DBMS and AI (especially expert systems) becomes obvious. Rules to deduce facts from other facts are

- a high-level specification of automatic reactions of a DBMS and
- an abstract, implicit definition of concrete, explicit actions.[100]

Therefore, deductive rules can be seen as a generalization of the concept of relational views[101]. The pragmatic purposes of deductive databases are[102]

- to accommodate simultaneous, but alternative views on a database, without redundant data replication,
- to support integrity checking in databases,
- to introduce additional terminology without adding redundant data,
- to augment or restructure databases, without moving data around in the store,
- to integrate heterogeneous data from different sources, and
- to conveniently express recursively defined data.

Deductive rules[103] may react to two different event types: query and update. Query events lead to the computation of derived data which is not stored in the database and may return answers to the query. Update events trigger the computation of data which is stored in the database (i.e. materialized views) and may include integrity checks. However, the rule execution in a deductive DBMS can

---

[97] Cf. Gallaire et al. (1984); Gallaire (1988); Harrison/Dietrich (1993); Quer/Olivé (1993); Karagiannis (1994), p. 22.

[98] Cf. Gallaire/Minker (1978).

[99] Cf. Gallaire et al. (1984), p. 169.

[100] Cf. Manthey (1993).

[101] Cf. Urban et al. (1992), p. 565.

[102] Cf. e.g. Nicolas/Yazdanian (1978); Azarmi (1992); Manthey (1993).

[103] For details on theses rules cf. e.g. Delcambre/Etheredge (1988); Widom/Finkelstein (1990); Hanson/Widom (1993).

lead to chains which are hard to control and validating this rule behavior is still an open research topic[104].

To complement a standard database with deductive rules, three generic approaches can be followed[105]:

- *Loose coupling*: This approach combines an existing database with an existing logic programming language. The two components are connected by an interface. This approach does not satisfy the objective of deductive databases and results in redundant rules and often poor performance.
- *Tight coupling*: Following this approach, the rule interpreter is modified or extended to support inference over database predicates. The interpreter is put as a layer upon the DBMS which allows for the optimization of query and rule processing.
- *Integration*: In this approach the inference subsystem is integrated into the DBMS; therefore, this approach demands for a modification of the DBMS itself. However, this allows for achieving an optimal coexistence and cooperation of deductive rules and the database.

Following one of those approaches, several prototypes of deductive databases have been developed or are still under construction[106]. Regarding the implementation of business rules, only those business rules which have the same scope as deductive rules, i.e., which prescribe the derivation of data. may be implemented in deductive DBMS. However, in order to prevent a too heterogeneous implementation, it is desirable that an implementation alternative covers more than a small subset of all rules.

2.2.2.3   Active Database Management Systems

2.2.2.3.1   Introduction

Conventional (passive) DBMS only serve as systems to store data in predefined data structures. With the exception of some integrity checking, e.g., in Oracle V6[107], the DBMS does not have the capability to perform any actions on its own. Active DBMS on the other hand, use rules based on the event-condition-action (ECA) paradigm to describe the activity in the DBMS[108]. Thus, the scope of da-

---

[104] Cf. e.g. Karadmice/Urban (1991).

[105] Cf. Zicari/Bauzer-Medeiros (1992), p. 151.

[106] For information on the research prototypes cf. Caseau (1991); Cacace et al. (1993); Ramakrishnan et al. (1993).

[107] Cf. Kosciuszko (1992a); Kosciuszko (1992b).

[108] Cf. Dayal (1988), p. 151.

tabases is enlarged in such a way that the DBMS monitors events and may react appropriately[109]. Active databases have been defined as „database systems that respond *automatically* to events generated internal or external to the system itself *without user intervention*"[110].

Early propositions for active mechanisms have been part of the CODASYL data description language[111] and since the first proposition of *database triggers*[112] as a functionality of DBMS, several research projects have been carried out to specify architecture and semantics[113] and to (partly) implement components of active DBMS. These prototypes make either use of a relational[114] or an object-oriented[115] DBMS.

Some results of the research in active DBMS are already incorporated in commercially available DBMS[116], e.g., Sybase (introduced database triggers already in 1986), Ingres[117] (1989), Oracle[118] (1994) and Illustra[119]. Furthermore, the SQL3 standard[120] will probably encompass a proposition for standardizing the currently divergent trigger syntax.

---

[109] For a comprehensive overview of the current state of the art in the field of active DBMS cf. Widom/Ceri (1996). In Jaeger/Freytag (1995), a bibiliography of this research field is given.

[110] Bauzer-Medeiros/Pfeffer (1991a).

[111] CODASYL (1973).

[112] Cf. Eswaran (1976), pp. 243.

[113] For an overview on these system cf. Chakravarthy (1993); Chakravarthy (1996).

[114] Cf. e. g. Stonebraker et al. (1987); Stonebraker et al. (1989); Haas et al. (1990); Widom et al. (1991); Hanson (1992); Stonebraker (1992). For an overview cf. Behrends (1995), pp. 46.

[115] For the fundamental concepts of OODBMS cf. Atkinson et al. (1989); Dittrich (1992). For the use of active mechanisms in OODBMS cf. Dayal (1988); Chakravarthy (1989); Bauzer-Medeiros/Pfeffer (1991b); Diaz et al. (1991); Anwar et al. (1992); Diaz/Embury (1992); Gehani et al. (1992); Kim et al. (1992); Branding et al. (1993); Lockemann/Walter (1993); Chakravarthy et al. (1994c); Kappel et al. (1994b); Urban et al. (1994). For an overview cf. Behrends (1995), pp. 50.

[116] For an evaluation the database triggers in Ingres, Oracle, and Sybase cf. Edelstein (1992); Schlesinger/Achermann (1995). The implementation of integrity constraints based on databased triggers is discussed in Neumann (1994). An overview on database servers is given e.g. in Khoshafian et al. (1992), pp. 22.

[117] For the application of event alerters as a component of rules in Ingres cf. Rennhackkamp (1992), pp. 86.

[118] The applicability of Oracle V7 database triggers has been investigated in Knolmayer/Schlesinger (1994a).

[119] Illustra is the commercial version of the research prototype Postgres.

[120] Cf. ANSI (1989), pp. 57; ANSI (1989), pp. 464; Melton (1993).

Active DBMS can be used for very different purposes[121] which can be separated in the support of applications based on the DBMS and in implementing and supporting internal functions of the DBMS. The first group encompasses among others network management, air traffic control, program trading, computer integrated manufacturing, engineering design[122], and office flow control[123]. Internal purposes are e.g. integrity control, access control, derived data handling, definition and application of inheritance mechanisms, alerting, performance measurement, support for inference, and the support of data interchange[124]. Especially the support of consistency enforcement is a largely stretched advantage of rules in DBMS, because the implementation of consistency constraints in the application programs leads to following problems[125]:

- *Larger source code*: Beneath the data manipulation all integrity constraints have to be implemented within the application code.
- *High degree of redundancy*: Different applications may check similar consistency constraints.
- *Difficulty of changing constraints*: Resulting from redundancy, the maintenance of constraints becomes difficult.
- *Inflexibility*: Most systems which provide mechanisms for consistency enforcement only cover common cases, i.e., exceptions are mostly not considered.
- *Unreliability*: Errors in the redundant decentralized consistency constraints are hard to discover.
- *Security*: Data entered interactively or imported from another database is not checked if the integrity constraints are only enforced by application programs.
- *Efficiency*: The application programmer often lacks the knowledge to efficiently implement the consistency constraints.

Therefore, a centralization of consistency enforcement in the DBMS is desirable and was proposed as early as 1975 by Eswaran and Chamberlin[126].

---

[121] Cf. Dittrich et al. (1986), p. 25; Dayal et al. (1988a), pp. 129; Chakravarthy et al. (1993); Fischer (1994).

[122] Cf. Kotz et al. (1988); Diaz et al. (1994); Urban et al. (1994).

[123] Kendler (1982); Kappel et al. (1994a).

[124] Cf. Dayal et al. (1988a), pp. 130. The applicability of database triggers in commercially available DBMS is described e.g. in Khoshafian et al. (1992), pp. 38; Knolmayer et al. (1994).

[125] Cf. Groff/Weinberg (1990), pp. 290; Eick/Werstein (1993), p. 52.

[126] Cf. Eswaran/Chamberlin (1975).

Beneath the basic ECA components, a rule in active DBMS may have additional properties[127] which encompass among others coupling modes[128], timing constraints, and contingency plans. Some of these additional properties are optional and complement the semantics specified by the basic ECA components.

2.2.2.3.2 Rule Syntax

Rules in active DBMS are mostly structured according to the ECA structure[129]. One problem in using ECA rules is the semantic distinction between event and condition. In certain cases, the assignment of a fact to either the event or the condition may be ambiguous. The event *'phone call of a customer between 8:00 a.m. and 5:00 p.m.'* as an example may also be formulated as an event *'phone call of a customer'* and the condition *'time between 8:00 a.m. and 5:00 p.m.'*[130]. This ambiguity may be circumvented by using different subtypes of ECA rules[131]:

- *Condition-Action* rules represent the AI approach to active rules (cf. Figure 2-1) and do not have an event component; therefore, they are triggered by a pattern matching in the database and an event like the one specified above would entirely become part of the condition.
- *Event-Action* rules have either more complicated event structures by incorporating the condition semantics in the event part or else assign the condition semantics to the action component.

Because of the practical importance of the complete ECA structure, we focus in the following on this type of active rules. For the specification of ECA rules, the event component is of special interest because of its rich semantic and complexity[132]. During the last five years especially the event algebras of Snoop and SAMOS have caused much interest; thus, these two algebras are presented in the following subsections.

---

[127] Cf. Dayal (1988), pp. 134; Comai et al. (1995), pp. 191.

[128] Cf. Hsu et al. (1988), pp. 173. The application of these coupling modes may lead to additional transaction problems as those described in McCarthy/Dayal (1989), pp. 217.

[129] Dayal (1988), p. 151.

[130] Cf. Herbst et al. (1994), p. 31.

[131] Cf. e.g. Dittrich et al. (1986), p. 29.

[132] Events in DBMS are further discussed e.g. in Reinert (1994).

2.2.2.3.2.1   Event Algebra of Snoop

Snoop is a specification language for active DBMS[133] which focuses on the event algebra and was developed at the University of Florida. In Snoop an event is defined as „an atomic [...] occurrence"[134]. In order to separate the conceptual and the physical level, a distinction between *logical events* and *physical/internal events* is made. Furthermore, events are classified in several primitive and composite event types (cf. Figure 2-14).

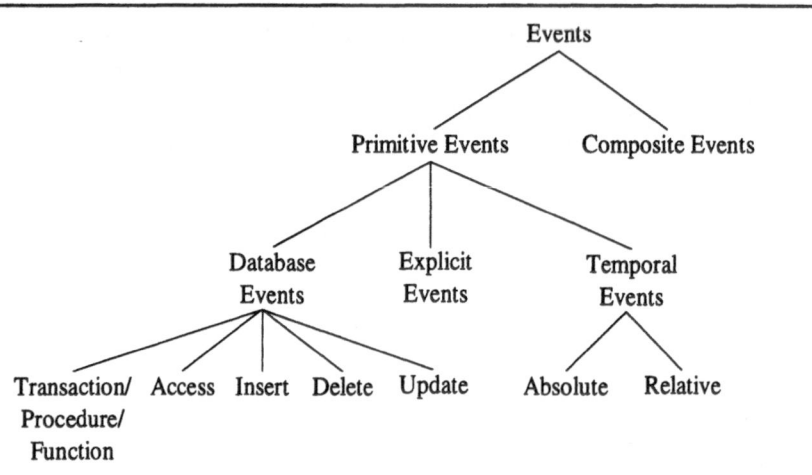

Figure 2-14: Event classification of Snoop[135]

Primitive events

Snoop distinguishes three main types of primitive events with different syntax:

- *Database events* encompass the four basic data manipulations access, insert, delete, and update of data stored in a database. Additionally, events related to the start or end of transactions, procedures, and functions are considered. The following examples specify the beginning of a transaction and the end of a delete operation as events[136]:

---

[133] For details on the event algebra of Snoop cf. Chakravarthy/Mishra (1993); Chakravarthy et al. (1994b).

[134] Chakravarthy/Mishra (1993), p. 4.

[135] Cf. Chakravarthy/Mishra (1993), p. 8.

[136] The examples are taken from Chakravarthy/Mishra (1993), p. 16. The syntax for primitive database events is not specified precisely; thus it can only be assumed from the examples given.

> ON   *(begin-of-transaction)*
> IF   ....
>
> ON   *(end-of-delete)*
> IF   ....

- *Explicit events* are events that are not part of the Snoop event language. They are defined by users or application programs. These events are expected to be detected outside of the system and then signaled to the system together with the accompanying parameters.
- *Temporal events* are specified by a point on the time line. In Snoop absolute temporal events and relative events are distinguished:
  - *Absolute temporal events* are specified with an absolute value represented as a time string: <(hh:mm:ss)mm/dd/yyyy>. The syntax explicitly allows for the use of wild cards, thus leading to events triggering a report at 5 p.m. each day:

    > ON   *(17:00:00)*/*/*
    > THEN   *print report*

  - *Relative events* correspond to a unique point on the time line but are specified as a delay starting from the occurrence of another event: event + [timestring].

Composite events

Composite events in *Snoop* are specified by the application of event operators on other events which may be either primitive or composite. These operators include disjunction, sequence, conjunction, and aperiodic and periodic event operators[137]:

- *Disjunction* ($\vee$): The disjunction of two events E_1 and E_2, specified as $E_1 \vee E_2$ occurs when at least one of the two events is signaled.
- *Sequence* (;): The sequence of two events E_1 and E_2 is denoted as $E_1;E_2$ are signaled when E_2 occurs after E_1.
- *Conjunction* (Any, All): The conjunction event is specified as Any(I, E_1, E_2, ..., E_n) where $I \leq n$. It occurs when any I events out of the n events are raised. The temporal order of the occurrence of the constituting events is irrelevant. The keyword 'All' is used as a shorthand for the case that $I = n$, i.e., all events of the list have to occur.
- *Aperiodic operator* (A, A^*): This event operator provides the specification of events bound by two arbitrary events for providing an interval: A(E_1, E_2, E_3). This event is signaled each time E_2 occurs within the specified interval.

---

[137] For examples cf. Chakravarthy/Mishra (1993), pp. 15.

The operator A^* is used to specify that the composite event is only signaled first time E_2 occurs within the interval. However, because the interval is defined by the occurrence of E_1 and E_3 this composite event could also be specified as a sequence with a negation of E_3: $(E_1; E_2; NOT E_3)$.

- *Periodic operator* (P, P^*): The periodic operator is used to repeat an event during a specified interval: $P(E_1 [timestring], E_3)$. This operator facilitates the specification of an event which triggers a report at the end of each quarter in 1996:

    P((00:00:00)01/01/96, [(00:00:00)03/00/00], (*.*.*)01/01/1997)

In the context of business rules, the Snoop event algebra serves among other things as a basis for the definition of the event syntax[138].

2.2.2.3.2.2   Event Algebra of SAMOS

The active DBMS SAMOS[139] was developed at the University of Zurich and is built on top of the object-oriented DBMS ObjectStore[140]. Similar to Snoop, the events provided by SAMOS are either primitive or composite[141] (cf. Figure 2-15); however, the types of primitive events and the operators for the specification of composite events differ from those of Snoop.

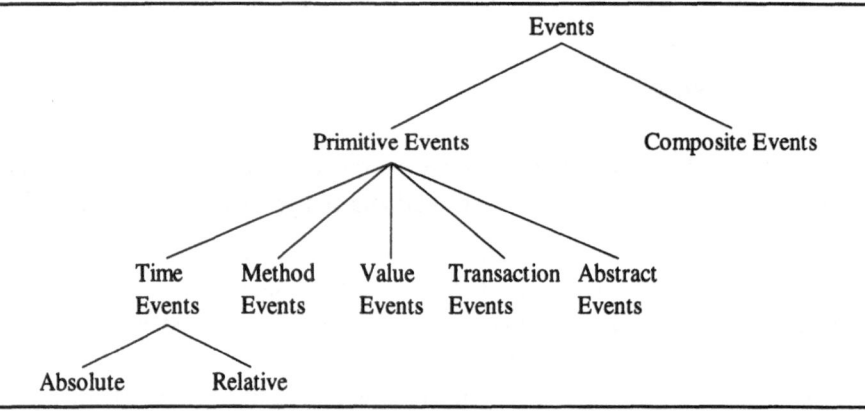

Figure 2-15: Event classification of SAMOS[142]

---

[138] For the business rule syntax cf. Chapter 3 and Appendix A.

[139] SAMOS is an acronym for **S**wiss **A**ctive **M**echanism Based **O**bject-Oriented Database **S**ystem.

[140] Cf. ObjectDesign (1991); Lamb (1991).

[141] All event types of SAMOS are described as BNF in Gatziu (1994), pp. 169.

[142] Cf. Gatziu (1994), pp. 43.

Primitive events

SAMOS provides five types of primitive events:

- *Time events*: Points in time are specified as absolute, implicit, or relative[143]:
  - *Absolute points in time* are specified by year, month, day, hour, and minute (wild cards are replaced e.g. by the current month) which leads to the specification of a single point in time. However, *repeated absolute time specifications* are supported, too (e.g. 'EVERY DAY 20:30').
  - *Implicit points in time* are not known at the moment of their specification but are related to a system activity.
  - *Relative points in time* are specified as (t+x) where *t* is a point in time and *x* is a time unit which is called relative time. The event 'Offer-sent + 30 DAYS' becomes concrete when a specific offer is sent (e.g. an offer sent on the '15.09.1996' leads to the event '15.09.1996 + 30 DAYS').
- *Method events*[144]: Method events result from the object-orientation of the underlying DBMS in which objects are accessed and manipulated by sending messages that invoke methods of the object. Method events can be related either to the start or the end of a method execution.
- *Value events*[145]: Though modifications of object values may only be done using methods, SAMOS supports an event type which is triggered by an update or read operation of an object value. This is appropriate because of the modularization of event specification: An update operation on an object of the class *Customer* may be used in several methods. Therefore, an event reacting to such an update would have to be specified redundantly for each method in which the update is executed.
- *Transaction events*[146]: The begin, end, or abort of a transaction may be relevant for following operations to be performed. The keywords used for the three situations are 'BOT', 'EOT', or 'ABORT' which leads to events like e.g. 'BOT payroll', where *payroll* is the name of a specific transaction.
- *Abstract events*[147]: Abstract events are named with a specific name and explicitly raised by a user or an application using the command RAISE EVENT [name].

---

[143] Cf. Gatziu (1994), pp. 38.
[144] Cf. Gatziu (1994), pp. 44.
[145] Cf. Gatziu (1994), pp. 45.
[146] Cf. Gatziu (1994), p. 46.
[147] Cf. Gatziu (1994), pp. 46.

Composite Events

Composite events in SAMOS[148] are divided in two groups, the first group encompasses event constructors with arity 2 (binary constructors which involve two other events) and the second group those with arity 1 (unary constructors which involve a single event).

- Binary constructors of SAMOS are conjunction (E_1,E_2), sequence $(E_1;E_2)$, and disjunction $(E_1|E_2)$. Because of the similarities to Snoop, these events are not further discussed.
- Unary constructors (arity 1)
    - ** and *last* constructor: The * constructor is used when an event is to be signaled only the first time it occurs within a defined interval. The * constructor therefore expresses similar semantics to the aperiodic operator of Snoop. The *last* constructor is used when the last occurrence of an event in an interval is to be signaled. However, such a last occurrence can not be signaled immediately but at the end of the interval specified.
    - *TIMES* constructor: The composite event specified as [TIMES (n, E) IN I] is signaled each time *n* events *E* occur within the time interval *I*. Regarding the event algebra of Snoop, the TIMES constructor is equivalent to its aperiodic operator.
    - *NOT* constructor: The event [NOT E in I] is signaled if the event *E* did not occur within the interval *I*.

2.2.2.3.3   Open Research Questions in Active DBMS

In the research area 'Active DBMS' several open questions and problems exist. On the technical level, they encompass the following:

- Should the active functionality be integrated in a DBMS or should it be part of an independent active layer[149] between the user applications and an (arbitrary) passive DBMS[150]?
- Are the event algebras proposed by different researchers sufficient for events relevant in practical applications, i.e., are they complete?
- How can the behavior of systems based on active DBMS be determined?
- How can inconsistencies and erroneous rule behavior be detected and solved?

---

[148] Cf. Gatziu (1994), pp. 47.

[149] As discussed e.g. in Kotz Dittrich (1992); Naqvi/Ibrahim (1993b).

[150] Cf. e.g. Naqvi/Ibrahim (1993a), pp. 58; Dayal (1995).

Beneath the three first aspects which are assigned to the physical/internal level, especially the *conceptual modeling of active mechanisms* is currently a main problem which has to be solved prior to a wide use of active DBMS in practice[151].

2.2.2.4 Practical Relevance of Deductive and Active Database Management Systems

The research on deductive databases started a long time ago and resulted in several fully developed prototypes; however, the only type of deductive rules provided by today's mostly used commercial DBMS are views. Deductive rules resulting from AI, which have been the focus of the research in deductive databases, do not seem to be a major demand of DBMS users. Similar to deductive rules, active rules resulting from database research (i.e. database triggers) are already incorporated by major relational DBMS and seem to match vital requirements of many DBMS users[152]. On the other hand, the implementation of production rules is still rather neglected by DBMS vendors.

The practical relevance of all technical developments described above must have the potential to more efficiently support the implementation of IS. From this point of view, it is remarkable that only very few research projects aim to apply results of the research in active and deductive DBMS in building practical IS. Such applications encompass

- cancer research[153],
- CIM systems[154], and
- workflow systems[155].

Such practical applications would finally close the feedback loop between theory and practice and may result in new problems to be solved.

---

[151] Cf. Herbst et al. (1994).

[152] Cf. Rytz (1994).

[153] Appelrath et al. (1993), pp. 74.

[154] Groiss/Eder (1993).

[155] Eder et al. (1994); Kappel et al. (1995).

2.2.3 Comparison of the Implementation Alternatives

For the implementation of business rules, five alternatives have been discussed which are put together in Figure 2-16. Though active DBMS are still not mature and often lack of important functionality, e.g., tools for rule administration, they can be considered as the most promising technique for a dynamic and modular implementation of business rules. Furthermore, practical applications are becoming an important research focus within the active DBMS area which will stimulate further improvements within the next few years.

	Implementation alternative				
	3GL	4GL	Rule-based 5GL	Deductive DBMS	**Active DBMS**
Events	No	Partially	Pattern	Pattern	**Yes**
Conditions	Branch	Branch	Yes	Yes	**Yes**
Actions	Yes	Yes	(Yes)	Yes	**Yes**
Special properties			Situation-Action rules	CA rules	**ECA rules**
Maturity	+++	++	++	+	+
Usability	-	+	+	++	+++

Figure 2-16: Evaluation of alternatives for the implementation of business rules

2.3 Summary

Rules are important elements of several *organizational theories*, e.g., scientific management, bureaucracy model, and BPR. However, there exist organizational approaches which emphasize the reduction of formalization. In the context of IS development, the organizational theory has the following impact on conceptual modeling and implementation of business rules.

For *specifying* a universe of discourse in a conceptual model, the subset of business rules to be formalized have to be selected. For this, different criteria for determining the appropriate level of formalization of business rules have to be considered. These criteria and their evaluation may result from the organizational approach which influences the members of the organization.

In our context, the *implementation* of IS is done in order to support the business of an organization. For the implementation of business rules in an IS, (1) the rules to be implemented have to be selected and (2) an appropriate implementation technique has to be chosen in order to achieve a flexible administration of the implemented rules. For the implementation of business rules in an IS, different alternatives may be considered. One of them is the decentralized implementation in program code using a 3GL or 4GL which often results in scattering business rules all over the IS. This may make their implementation and maintenance rather difficult and may lead to inconsistencies. Another alternative is using rule constructs provided by deductive and active DBMS. It has been shown that though active DBMS are not available in their full functionality they will be suited for a modular implementation of business rules and their event-oriented processing.

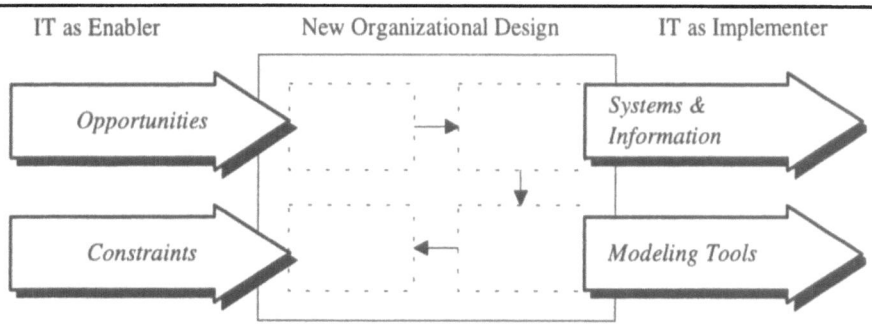

Figure 2-17: The role of IT in organizational design[156]

---

[156] Davenport (1993), p. 49.

The interdependencies between IT and organizational design have been discussed by Davenport (cf. Figure 2-17). In this context, IT does not only provide constraints and opportunities for organizational design but also serves as an implementer. Thus, the discussion of topics of the organizational design results in requirements for (1) methodologies of systems and information engineering and (2) modeling tools. In our context, the following main requirements for a *business rule-oriented conceptual modeling* have to be considered:

- In order to maintain a large set of business rules, a classification system has to be provided which is based on related organizational facts.

- To support the communication of the resulting model to the members of an organization, the specification of business rules should be based on a syntax which is precise and can easily be understood by non-trained end users. Furthermore, different views on the whole model should be provided (e.g. process view, data structure view, behavior view, and several specific organizational views).

- Because of an assumed orientation towards organizational dynamics, the main properties of business processes should be specifiable. However, because of using business rules as a new structure in the conceptual model, the current goal is to evaluate the applicability of business rules on the specification of the dynamic properties of processes, i.e., the specification of processes on different abstraction levels.

In the following chapter, a case study on business rules, and a classification system for integrity constraints and business rules will be presented. Furthermore, the relationships between business rules and closely related facts are discussed. Finally, the currently only sketched syntax for the specification of different types of business rules is specified.

3 BUSINESS RULES

IS normally encompass a large amount of business rules; the administration of these rules in a conceptual modeling is only feasible if the rules can be regarded and classified with respect to other facts:

- Business rules should be analyzable with respect to organizational facts.
- Business rules should be analyzable with regard to their validity, i.e., their correctness with respect to the real world and with respect to their mutual consistency.
- Depending on the rule type, the syntax for the specification of business rules has to be more or less restricting.
- Different business rule types may be assigned to specific implementation alternatives, e.g., those discussed in Chapter 2.

In the preceding chapter the formalization of rules in organizational theory and selected alternatives for their implementation in IS has been discussed. This chapter aims to provide the foundations for the meta model and the modeling steps of *BROCOM*.

First, case studies on the relevance of business rules in practically applied IS and examples of business rules are presented. One main problem of administrating business rules in practical IS is their large number. The resulting complexity can be reduced by e.g. distinguishing different rule types; therefore, a system for the classification of business rule components is introduced which is directly used as a basis for a syntax for business rules. Because the conceptual modeling of business rules must not be confined to business rules themselves, their organizational environment is taken into account. This environment encompasses processes, processors, organizational units, and origins and the relationships between them.

3.1 Case Studies

3.1.1 Overview

In our research project, three case studies on business rules have been carried out[1]. The main goal of the case studies was to gather information on business rules relevant within real IS (cf. Figure 3-1) [2]. Though the overall goal of all three studies was the same, each of them served a specific purpose:

- *DAZ[3]:* The case study DAZ[4] deals with an application of the Swiss Bank Corporation. The analyzed IS covers a part of the decentral payment administration. This was the first study and provided us with practical information on business rules and gave first insights into the problem field.
- *BALICO[5]:* In the case study BALICO[6] an IS dealing with the administration of real estates of the Swiss PTT was analyzed. The study attempted (1) to analyze types of business rules relevant in an IS and (2) to evaluate the possibilities of transforming them from applications written in Oracle*Forms 3 to database triggers of the DBMS Oracle 7.0. For this purpose, rules found in the program code were marked, examined, and classified.
- *DOM-2[7]:* This case study deals with the damage administration system of the Swiss Mobiliar insurance company. The specific goal of the case study DOM-2[8] was to provide a large set of rules as a basis for the practical application of the **BROCOM** approach presented in the next chapter. While in the first two case studies, rules were just analyzed within the program code, business rules found in DOM-2 were been collected and registered in a database which eventually contained over 750 business rules.

---

[1]   Cf. Knolmayer et al. (1994), pp. 27.

[2]   In Bell (1990b), another case study on business rules in relational databases is presented. A methodology for the extraction of business rules from procedural code is discussed in Glasier (1992) and Hanks (1992); however, the scope of business rules in this methodology remains rather unclear.

[3]   DAZ is the acronym for 'Dezentral Auftragserfassung Zahlungsverkehr' (decentral registration of payments).

[4]   Cf. Eisenegger (1994).

[5]   BALICO is the acronym for '**Bau und Liegenschaftswesen mit Computer**', (Computer supported administration of real estates).

[6]   Cf. Knolmayer/Schlesinger (1994b).

[7]   DOM is the acronym for the French word **dom**age (damage).

[8]   Cf. Herbst (1996).

	„DAZ"	„BALICO"	„DOM-2"
Enterprise	Swiss bank company	Swiss PTT	Swiss insurance company
Purpose of the IS	Decentral entry of payments	Administration of real estates	Administration of damage claims
Development environment	COBOL, IMS	Oracle*Forms 3, Oracle V6	COBOL, IMS
Research Methods	Code analysis, interviews	Code analysis	Code analysis, interviews
Lines of code (approx.)	1'500	55'000	85'000
Number of business rules	76 (100 %)	523 (100 %)	758 (100 %)
... in program code	64 (84 %)	523 (100 %)	627 (83 %)
... in system environment	12 (16 %)	0 (0 %)	131 (17 %)

Figure 3-1:   Overview of the case studies

The three case studies revealed the large number of business rules in a real IS as the major problem[9]. Therefore, the *BROCOM* approach has to provide possibilities to reduce this complexity. In order to achieve this, business rules are

1. classified,
2. formalized separately for each rule class, and
3. specified in a top-down approach on different abstraction levels.

The classification of business rules is discussed in Section 3.2 and the specification aspects are subject of Section 4.2 which deals with the steps of the *BROCOM* approach.

3.1.2   Case Study 'DOM-2'

The case study DOM-2 was started in 1992 with first interviews in the information systems department of the Swiss Mobiliar insurance company. The IS 'DOM' consists of several subsystems. One of them is DOM-2 which supports the processing of insured events, i.e., damages and covers the fields personal liability insurance, third party insurance, insurance against theft, and other property insurance. In the case study two different aspects were considered:

---

[9] Furthermore, the case study resulted in shortcomings of existent active components in DBMS with respect to the phenomena experienced and therefore allows to draw some requirements for those mechanisms. For details cf. Herbst/Myrach (1996).

- *Business process:* The business process of the damage administration was analyzed in interviews. The focus of the interviews was the process dynamics; other properties of business processes such as process goals, consumed resources, and added values were neglected. The interviews resulted in 131 business rules which define the business process as exact as possible and desirable.
- *Terms of insurance policies:* The terms of insurance policies describe whether a damage is covered by a certain policy. These terms are specified in the policy contracts, accompanying documents, and program documentation and are of course embedded in program code of the damage administration IS. The documents provided by the industrial partner encompass

  - *verbal descriptions* of the insurance terms,
  - *functional models* of the IS,
  - the whole *program code* (approximately 80'000 to 90'000 lines of COBOL code including data and procedure divisions),
  - *screen descriptions*, and
  - the *database schema* of the hierarchical IMS structures.

  The analysis of these documents resulted in 627 rules like the following:

  ON   *((damage-field entered) OR (damage-cause entered))*
  IF   *(damage-field = private third party insurance) AND*
          (damage-cause = damage of a car in use) AND
          (third-party-insurance-type = family, single or senior)
  THEN   *issue error message „Damages of cars in use are not covered by this policy; please check and, if necessary, pass the file to the central office."*

This example illustrates the difficulty of unambiguously separating business rules for integrity constraints and for process specification: The negative evaluation of the integrity condition of the example rule results in an error and may lead to passing the file to the central office; this fact is certainly relevant for the business process and must not be excluded from its specification.

3.1.3 Examples of Business Rules

To illustrate the scope and different types of business rules, a set of rules from the case study DOM-2 is introduced which describes the business process *damage reception*. In the rule specification, words written in capitals represent names of events or rules (e.g. the event which triggers the second rule is *(damage reported)*; this event is raised in the first rule where the name of the event, i.e., 'DAMAGE-REPORTED' is used). The names of events are also used in the event specification to compose complex events, e.g., the event of the first rule.

The process is initiated by any person getting in contact with the insurance company. The first business rule encompasses a check, if the person is a policy-holder of their company and if the policy-holder wants to report a damage.

Business rule [1] 'PERSON-CONTACTS-US'
ON   (PHONE-CALL-OF-PERSON) OR (LETTER-OF-PERSON)
IF   (person is a policy-holder) AND (policy-holder reports damage)
THEN   begin damage registration;
*   raise event 'DAMAGE-REPORTED'*

The execution of this business rule leads to the provisional registration of the damage if the policy-holder already provides information on the damage; in the other case he receives a form to fill in.

Business rule [2] 'REGISTER-PROV-DAMAGE'
ON   (damage reported)
IF   (information about damage available)
THEN   register damage provisionally;
*   registration-date := TODAY ();*
*   state := provisional;*
*   raise event 'DAMAGE-PROV-REGISTERED'*
ELSE   send damage form to policy-holder;
*   raise event 'DAMAGE-FORM-SENT'*

The registration of the incomplete information about the damage is followed by checking whether the damage is covered by a policy of the policy-holder. Depending on the result, either the claims of the policy-holder are rejected or a form to complete the information about the damage is sent to the policy-holder.

Business rule [3] 'ACCEPT-PROV-DAMAGE'
ON   (provisional damage registered)
IF   (damage covered by policy)
THEN   accept damage provisionally;
*   raise event 'DAMAGE-PROV-ACCEPTED';*
*   send damage form to policy-holder;*
*   raise event 'DAMAGE-FORM-SENT'*
ELSE   reject damage;
*   raise event 'DAMAGE-REJECTED'*

The policy-holder is reminded if he does not return the form within 60 days.

Business rule [4] 'REMINDER-FOR-DAMAGE-FORM'
ON   (60 days after (DAMAGE-FORM-SENT))
IF   (damage form not returned)
THEN   remind policy-holder

Another 30 days later, the provisionally registered damage is deleted if the form has not been returned.

> *Business rule [5] 'TIME-OUT-FOR-DAMAGE-FORM'*
> *ON   (90 days after (DAMAGE-FORM-SENT))*
> *IF   (damage form not returned)*
> *THEN   delete provisionally accepted damage*

The last rule in the example is triggered by the events *'damage form sent'* and *'damage form received'* which have to occur in conjunction. Every received form is checked with regard to its completeness and returned to the policy-holder if it is incomplete. Otherwise the definitive registration of the damage can be started.

> *Business rule [6] 'DEF-REGISTER-DAMAGE'*
> *ON   (DAMAGE-FORM-RECEIVED WITHIN (90 days after*
> *   (DAMAGE-FORM-SENT)))*
> *IF   (information is complete)*
> *THEN   raise event 'START-DEF-DAMAGE-REGISTRATION'*
> *ELSE   return form to policy-holder;*
> *   raise event 'DAMAGE-FORM-SENT'*

Subsequent to the execution of these business rules, the process of treating the definitive damage claims of the policy-holder may begin.

3.2 Classification of Integrity Constraints and Business Rules

The definition of a classification system should strive for a clear systematic, i.e., a precise formulation of the classification criteria, which guarantees the completeness and uniqueness of the classification and selection of facts. First , this allows the classification of all kind of business rules to specific classes and second, an easier retrieval of rules (e.g. for rule maintenance) by certain selection criteria[10]. Another goal of a classification system is the orthogonality of its criteria.

On a very general level, business rules can be distinguished into two classes[11]:

- *Integrity rules* describe allowed states and state transitions of data that is stored in a database or exists in the real world. Violating the condition of an integrity rule is semantically an error which may result in a (re)action. Integrity rules represent semantics of integrity constraints which are extended

---

[10] Cf. e.g. Hildebrand (1991), p. 15.

[11] Cf. Hsu/Cheatham (1988), pp. 150; in McBrien et al. (1991), p. 315, derivation rules as a third main class are discussed.

by making the triggering event and the reaction on a constraint violation explicit.

Example:

> ON   *(update stock of product)*
> IF   *(new stock < 0)*
> THEN *issue error message "The stock must not be lower than zero"*

- *Automation rules* describe the logic of a task execution; thus, the violation of the condition does not have the semantic of an error. In case of a rule with a *then-* and an *else-action*, a selection between the two mutually exclusive actions is described, whereas if only a *then-action* exists, nothing happens if the condition is not satisfied.

Example:

> ON   *(update stock of product)*
> IF   *(new stock < reorder-threshold)*
> THEN *reorder product*

The classification criteria for business rules are presented in two parts. The first encompasses a classification of integrity constraints as a special subset of all business rules and the second a classification of business rule components.

3.2.1   Types of Integrity Constraints

A database may contain any data, but not every state, i.e., situation of the database is valid in the context of an application for which it is used. Therefore, constraints on the possible database states are necessary in order to have only valid states and to assure the correct operation of each virtual machine[12]. In the following, selected criteria for the classification of integrity constraints are discussed which are relevant in the context of business rules[13].

3.2.1.1   Number of Database States Involved

The evaluation of an integrity constraint can consider either the actual state of the database or compare several states. This leads to the distinction of static and dynamic integrity constraints[14].

---

[12] Cf. Shepherd/Kerschberg (1986), p. 310.

[13] Because this is not a main focus of this work, the discussion is kept rather short. For details refer to the literature indicated. For further classification approaches, cf. e.g. Rebsamen (1983); Steinbauer (1983); Gähler (1987), pp. 13; Leikauf (1989); Elmasri/Navathe (1994), pp. 638; Herbst/Knolmayer (1994).

[14] Cf. Hammer/McLeod (1976), pp. 499; Wedekind (1976), p. 283; Ehrich et al. (1984), p. 301; Azarmi (1992), p. 28; Thalheim (1992), pp. 5; Lipeck (1992), pp. 43; Gertz/Lipeck (1993), pp. 23; Dietz (1994), p. 724; Theodoulidis et al. (1994), pp. 191.

- *Static integrity constraints*: For the evaluation of a static integrity constraint, only the current state has to be considered.

  The date of birth of a person has to be today or in the past

ON	*(insert person)*
IF	*(date of birth > current date)*
THEN	*error*

  Only a finite set of marital status of a person are valid:

ON	*(insert person)*
IF	*(marital status NOT IN ('not married', 'married', ' divorced', 'widowed'))*
THEN	*error*

- *Dynamic integrity constraints* specify allowed changes from state$_1$ to state$_2$. For the evaluation the old and the new state have to be regarded.

  Example: The following state changes of the marital status of a person are allowed:

not married	⇨ married
married	⇨ **divorced or widowed**
divorced	⇨ married
widowed	⇨ married

ON	*(update marital status of person)*
IF	*(marital status ('married') NOT TO ('divorced', 'widowed'))*
THEN	*error*

The distinction between static and dynamic constraints serves among others as a basis for the condition syntax which is already sketched in the examples given above and will be further discussed in Section 3.4.

3.2.1.2   Number of Referred Database Constructs

The evaluation of an integrity constraint may involve different amount and types of database constructs like attributes, tuples, and relations for a relational DBMS[15]. The fewer constructs are affected, the easier the constraint evaluation becomes. Bertino and Apuzzo distinguish

- tuple constraints,
- relation constraints, and
- multi-relation constraints[16].

---

[15] Cf. Eswaran/Chamberlin (1975), p. 54; Weber et al. (1983), p. 126; Bertino/Apuzzo (1984), p. 44; Reuter (1987), p. 381; Knolmayer/Herbst (1993), p. 387.

[16] Bertino/Apuzzo (1984), p. 44.

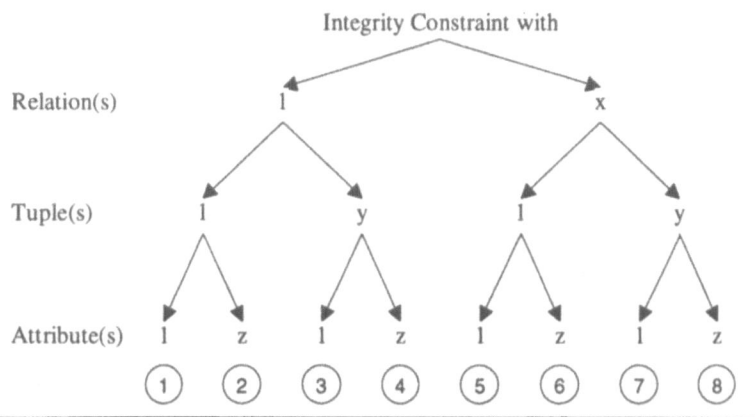

Figure 3-2:   Classification of integrity constraints with respect to referred constructs[17]

A similar classification can be done with respect to domains, objects, or relationships between objects[18].

A comprehensive overview of these constraint classes is given in Figure 3-2. Examples for the eight types are[19]:

- One relation (intra-relation constraints):

1. Domain constraints: NOT NULL constraint for an attribute which is part of a primary key.
2. Comparison of at least two attribute values of a single tuple: 'Begin-date of project $\leq$ End-date of project'.
3. Evaluation involves one attribute of several tuples: in a database which supports versions, the state change of an attribute results in comparing attribute values of several tuples, e.g., the constraint 'salary of old version < salary of new version'. Another example would be a uniqueness constraint in case of a combined primary key.
4. Constraint encompasses at least two attributes of several tuples: uniqueness constraint for composed primary keys.

- Several relations (inter-relation constraints):

5. Constraint involves a single tuple and attribute: referential integrity[20] if the foreign key consists of one attribute only.
6. Evaluation concerns a single tuple and several attributes: referential integrity if the foreign key consists of at least two attributes.

---

[17] Knolmayer/Herbst (1993), p. 387.
[18] Cf. Hammer/McLeod (1975), p. 29; Codd (1979), p. 400; Codd (1990), p. 246.
[19] Further examples are given in Date (1993), pp. 15.
[20] Cf. Date (1981), p. 3.

7. Constraint encompasses several tuples and one attribute each: cardinality constraints[21] involving complex cardinalities.
8. Comparison of values stored in several tuples and at least two attributes: temporal referential integrity constraints in a time extended data model[22] or a comparison of aggregated values, e.g., 'order total ≤ credit limit'.

3.2.1.3 Time Point of Evaluation

The decision on when to evaluate an integrity constraint depends among others on the kind of application in which it is used. In design applications, often temporary consistency violations have to be allowed[23]. Thus, the time point of the evaluation of a constraint may be[24]

- immediate (e.g. for the domain of an attribute),
- deferred (e.g. referential integrity constraints at the end of a transaction), or
- user initiated (e.g. when a CAD user decides that the design result has reached a state in which a consistency check is appropriate).

In the context of business rules, this classification allows for the determination of the event which triggers the evaluation of an integrity constraint.

3.2.1.4 Relation to the Data Model

With respect to the data model, integrity constraints are classified as inherent, implicit, or explicit[25]:

- *Inherent integrity constraints* do not have to be specified in a schema because they relate to the basic assumptions of the data model (e.g. the atomarity of an attribute value in the relational model).
- *Implicit integrity constraints* are defined by the elements of the data model used; examples for implicit integrity constraints in the relational model are data types, uniqueness of the primary key, referential integrity, relation cardinalities, and generalization rules.[26]
- *Explicit integrity constraints* are additional constraints resulting from the context of an application (e.g. allowed state transitions).

---

[21] Cf. Little et al. (1993), pp. 235.
[22] Cf. Myrach et al. (1996).
[23] Cf. e.g. Buchmann/Dayal (1988), p. 96.
[24] Cf. Eswaran/Chamberlin (1975), pp. 55; Lafue (1982), pp. 295; Weber et al. (1983), p. 126; Steinbauer/Wedekind (1985), p. 61; Reuter (1987), pp. 381; Date (1993), pp. 18; Eick/Werstein (1993), p. 53.
[25] Cf. Rebsamen (1983); Shepherd/Kerschberg (1986), p. 311; Elmasri/Navathe (1994), pp. 641.
[26] Leikauf (1990), p. 31.

In the *BROCOM* approach presented in the next chapter, one step encompasses the specification of integrity constraints. The classification in inherent, implicit, and explicit integrity constraints is used to subdivide the task[27].

3.2.1.5 Type of (Re)Action

The violation of an integrity constraint may result in two basic types of (re)actions[28]:

- *Correction*: In specific situations, an incorrect state of a database may be corrected by predefined actions. The SQL standard provides two ways of *correcting* violated referential integrity constraints[29]: SET NULL/SET DEFAULT or CASCADE.

  Example: Cascading deletion of orders which depend on a customer:
  ON   *(delete customer)*
  IF   *(exists order related to customer)*
  THEN   *delete related orders*

- *Rejection*: In other cases, the potentially corrupted state of a database cannot be automatically corrected. In the SQL standard the REJECT option leads to the abortion of each operation which violates an integrity constraint.

  Example: Abort deletion of a customer if related orders exist:
  ON   *(delete customer)*
  IF   *(exists order related to customer)*
  THEN   *error, reject deletion*

3.2.1.6 Database Schema

With respect to the life cycle of an object, integrity constraints can be classified into constitutive and regulative integrity constraints[30].

- *Constitutive integrity constraints* encompass constraints which have to be respected for the creation of a new instance of an object. The violation of a constitutive integrity constraint has always to be corrected because an object cannot exist while a constraint of this type is violated. As will be shown later, these integrity constraints are always static, because no older states exist.

---

[27] Cf. modeling step *integrity constraints* in Section 4.2.
[28] Cf. Hammer/McLeod (1975), pp. 30; Rebsamen (1983), pp. 53; Gähler (1991), p. 38.
[29] Cf. e.g. Melton/Simon (1993), pp. 221.
[30] Cf. Wedekind (1983), p. 134; Leikauf (1990), pp. 30.

Example: An object 'person' must have a valid date of birth
    ON   *(insert person)*
    IF   *(date of birth > current date)*
    THEN   *error*

- *Regulative integrity constraints* define the correct use of an existing object instance. The evaluation of these constraints normally succeeds the evaluation of constitutive constraints because the correct use of an object presupposes its correct existence.

Example: The salary of an employee must not be reduced
    ON   *(update salary of person)*
    IF   *(old salary > new salary)*
    THEN   *error*

3.2.1.7   Interdependencies between the Classification Criteria

The classification criteria presented above are not fully independent; the following interdependencies of the criteria discussed exist[31]:

- 'Number of states' vs. 'Relation to the data model': Inherent and implicit constraints are always static.
- 'Number of states' vs. 'Relation to the database schema': Constitutive constraints are always static.
- 'Time point of evaluation' vs. 'Relation to the database schema': Constitutive constraints always have to be checked immediately.

3.2.2   Types of Business Rules

The classification of the three basic rule components directly leads to a syntax for each of them. In order to enable an automated analysis of the rule content, a more or less restricting syntax of different component types is necessary[32].

3.2.2.1   Event Types

Events which trigger business rules can be classified according to their complexity and their content (cf. Figure 3-3). For each resulting event type, a syntax for its specification is presented[33]. On a first level, events are classified into *elementary events* and *composite events*[34]. The first are atomic and cannot be further

---

[31] Cf. Leikauf (1990), p. 36.
[32] A detailed specification of the syntax is given in Appendix A.
[33] Other classifications of events in systems analysis can be found e.g. in Lingat et al. (1987); Hoffmann et al. (1992); Teisseire et al. (1994).
[34] Cf. Chakravarthy/Mishra (1993), pp. 3.

decomposed, i.e., they may not encompass other events as constituting elements. Composite events are specified by using operators on other (elementary or composite) events.

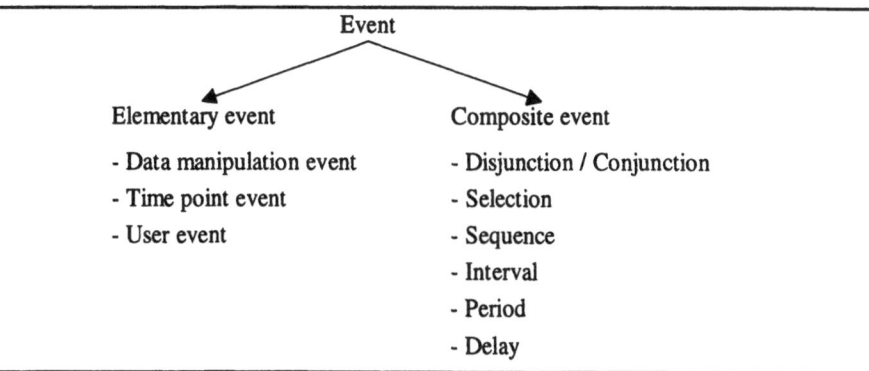

Figure 3-3: Event types

Elementary events can be classified into three major subclasses[35]:

- *Data manipulation events* are raised by manipulations of data objects; these manipulations encompass the generation of a new instance, the modification, the retrieval, and the deletion of an existing instance and the derivation of data:

  Syntax:

      ON   *(operation object)*

  Examples:

      ON   *(insert customer)*
      IF   ...
      THEN ...

      ON   *(update customer)*
      IF   ...
      THEN ...

      ON   *(update name of customer)*
      IF   ...
      THEN ...

      ON   *(delete customer)*
      IF   ...
      THEN ...

---

[35] A similar classification of elementary events is described in Souveyet/Rolland (1990), pp. 155; Skidmore et al. (1992), p. 175.

- *Time point events* are specified by an exact date and time and are raised when the clock is equal to this time point.

  Syntax:
  ON   *(mm.dd.yyyy/hh:mm:ss)*

  Example:
  ON   *(15.09.1996/12:00:00)*
  IF   ...
  THEN   ...

- *User events* do not directly result from a data manipulation and can only be detected and raised by users or monitor programs in applications. Because user events are rather unstructured and may encompass a large variety of events, a syntax for the event content would be too restrictive for the event specification and is therefore not desirable.

  Syntax:
  ON   *(any meaningful event description)*

  Example:
  ON   *(phone call of customer)*
  IF   ...
  THEN   ...

Composite events are specified by the application of operators to other events. According to the event algebras developed for rules in active databases[36], seven different types of composite events are distinguished:

- A composite event defined by a *disjunction* of events occurs when at least one of the constituting events is raised.

  Syntax:
  ON   *(Event$_1$) OR (Event$_2$)*

  Example:
  ON   *(phone call of customer) OR (letter of customer)*
  IF   ...
  THEN   ...

- An event specified by the *conjunction* of events happens when all constituting events are raised. This composite event type does not imply any temporal dependencies, i.e., the order of raising the single events is not implied.

---

[36] For event algebras cf. Chapter 2. Cf. also Chakravarthy/Mishra (1993), pp. 10; Gatziu/Dittrich (1993), pp. 25; Gatziu (1994), pp. 37.

Syntax:
ON   *(Event₁) AND (Event₂)*

Example:
ON   **(Offer sent) AND (Phone call of customer)**
IF   ...
THEN   ...

- A *selection event* is raised when a specified number m of a list of n events has occurred , whereas the sequence of the events is not relevant.

  Syntax:
  ON   *m of ((Event₁); (Event₂); ...; (Eventₙ))*

  Example:
  ON   **1 OF ((phone call of customer); (letter from customer))**
  IF   ...
  THEN   ...

Selection events can be considered as specification alternatives for disjunctions and conjunctions of events: If m is 1 the selection event is equivalent to a disjunction of all constituting events (cf. example above) and if m equals n the selection expresses a conjunction.

- A *sequence event* happens when all constituting events occur exactly in the specified chronological order. A sequence event is therefore a temporally ordered conjunction of events.

  Syntax:
  ON   *((Event₁) ; (Event₂) ; ... ; (Eventₙ))*

  Example:
  ON   **((offer sent); (offer rejected))**
  IF   ...
  THEN   ...

- An *interval event* is raised when a specified event E is raised within an interval which is defined by two other events.

  Syntax:
  ON   *(Event) WITHIN ((Event₁); (Event₂))*

  Example:
  ON   **(phone call of customer) WITHIN ((offer sent);**
          (30 DAY AFTER (offer sent)))
  IF   ...
  THEN   ...

The end of the interval has to be either a time point event or a delay event. Other event types defining the end of an interval may lead to impossible event detections:

Example:

ON   *(phone call of customer) WITHIN ((offer sent);*
        (offer returned))

IF   ...

THEN   ...

If an event is specified like this, it is only raised when the offer is finally returned. Prior to this it is not clear whether this event determining the end of the interval occurs at all. As an alternative the event could be specified as a sequential event with a negation:

Example:

ON   *((offer sent); (phone call of customer); NOT (offer returned))*

IF   ...

THEN   ...

- A *periodical event*[37] is raised when a specific event (e.g. 'offer sent') or a temporal specification (e.g. '1st day of a month at 6 p.m.') happens for the n-th time. In order to specify temporal periodical events the use of place-holders is allowed.

  Syntax:

  ON   EACH n (Event)

  Examples:

  ON   *EACH 10 (Offer sent)*

  IF   ...

  THEN   ...

  ON   *EACH 2 (01.*.*/18:00:00)*

  IF   ...

  THEN   ...

- A *delay event* happens after a duration d which begins at the time point of the occurrence of an event E. The time unit may be a second, minute, hour, day, week, month, or year.

  Syntax:

  ON   d Unit AFTER (Event)

---

[37] The term *period* can be used for a duration with a specified beginning and ending or can be defined as an event which happens regularly. In our context its second meaning is used.

Example:
> ON   *30 DAY AFTER (offer sent)*
> IF   ...
> THEN   ...

The classification of events discussed above relates to the conceptual level of rule specification. In order to have an unambiguous transformation of the conceptual specification to e.g. events of active DBMS, each event type on the conceptual level should be equivalent to exactly one event type of the active DBMS chosen. In Chapter 2 the event algebras of Snoop and SAMOS were introduced. In Figure 3-4 the event types of business rules are assigned to those of the two event algebras for the physical level.

| | Event types in | |
Business rules	Snoop	SAMOS
Time point event (p)	Absolute temporal event (p)	Absolute time (p)
Database event (p)	Database event (p)	Value event (p)
User event (p)	Explicit event (p)	Abstract event (p)
Disjunction/ Conjunction (c)	OR or ALL operator (c)	Disjunction/ Conjunction (c)
Sequence (c)	Sequence operator (c)	Sequence constructor (c)
Selection (c)	ANY operator (c)	- - -
Interval (c)	Aperiodic and periodic operator (c)	* constructor (c)
Period (c)	ANY operator (c)	TIMES constructor (c)
Delay (c)	Relative temporal event (p)	Relative time (p)

Figure 3-4:   Comparison of event types in business rules, Snoop, and SAMOS. Events marked with (p) are primitive and those with (c) are composed.

There exist interdependencies between the event types introduced above and the basic rule types, integrity and automation rules. Figure 3-5 puts together which type of event may trigger a certain type of business rule. Data manipulation events may trigger every kind of rule, whereas time point events are not appropriate to trigger the execution of business rules representing static or dynamic integrity constraints. User events however may trigger every kind of business rules, i.e., also integrity rules. In this case the user event could be e.g. a user defined integrity checking in a design application as discussed in Section 3.2.1.3.

Event type	Static integrity rule	Dynamic integrity rule	Automation rule
Data manipulation event	Yes	Yes	Yes
Time point event	No	No	Yes
User event	Yes	Yes	Yes

Figure 3-5:   Interdependencies of event type and rule type

3.2.2.2   Condition Types

Conditions of *integrity rules* can be classified with respect to the criteria discussed in section 3.2.1. Regarding all business rules, i.e., integrity and automation rules, the condition component can be classified on a first level into elementary and composite conditions. *Elementary conditions* are either conditions on sets or predicates[38] and have to be atomic, i.e., they may only encompass unary Boolean operators like NOT, but not the binary Boolean operators *AND* or *OR*.

A set condition is a query on the membership of an element in a set. They may check if a referred tuple exists (e.g. 'exist policy-holder related to policy') or if a value is an element of a finite set of values (e.g. 'exist policy-type IN {list of policy types}'). Predicates are comparisons of attribute values with other attribute values (e.g. 'stock < reorder-threshold') or constants (e.g. 'stock < 100').

Composite conditions result from the application of the binary Boolean operators *AND* or *OR* to elementary or composite conditions.

The syntax for the condition component has to allow an automated analysis of the condition content, e.g., the derivation of life cycles. Therefore, for the syntax of *elementary conditions*, three rule types are distinguished: automation and dynamic and static integrity rules.

Condition Syntax: Automation Rules

Automation rules do not result in errors depending on the evaluation of a condition but determine the next action to be executed. Thus the condition of automation rules may encompass a large variety of conditions. In order to have a flexible and not too restricting syntax, conditions of automation rules are not further constrained, i.e., they may contain any text.

> *ON   ...*
> *IF   <any text>*
> *THEN   ...*

---

[38]   Cf. Hanson/Widom (1993), pp. 6.

Condition Syntax: Integrity Rules for Static Integrity Constraints

Business rules which check static integrity constraints only refer to values of a specific time point. As discussed in Section 3.2.2.2, a condition may be either a value comparison or a set condition. For the specification of comparisons between a property and a value or another property, respectively, the following syntax is applied:

Syntax:
>*ON* ·   ...
>*IF*   (*<property> <Comparison> <list of allowed values>*)
>*THEN*   ...

or

>*ON*   ...
>*IF*   (*<property> <Comparison> <property>)*
>*THEN*   ...

Based on this syntax, the *reorder-threshold* condition already used before is specified as follows:

Example:
>*ON*   *(UPDATE stock of product)*
>*IF*   **(stock of product < reorder-threshold)**
>*THEN*   *reorder product*

The other type of elementary conditions check if a value is a member of a set.

Syntax:
>*ON*   ...
>*IF*   *(exist <property> IN (<list of allowed values>))*
>*THEN*   ...

As an example the value of the property *marital status* is regarded: The marital status has to be one of those specified in a finite set encompassing 'not married', 'married', 'divorced', and 'widowed'.

The static constraint for marital status may be formalized as follows:

>*ON*   *(update marital status of person)* .
>*IF*   *(exist marital status of person IN ('not married', 'married',*
>   *'divorced', 'widowed'))*
>*THEN*   *commit operation*
>*ELSE*   *reject operation*

Condition Syntax: Business Rules for Dynamic Integrity Constraints

Dynamic integrity constraints prescribe permissible state changes and make reference to an old and a new value of an object property. As an example the marital status of a person may have the following state changes:

- *not married*   ⇨ *married*
- *married*   ⇨ *divorced or widowed*
- *divorced*   ⇨ *married*
- *widowed*   ⇨ *married*

Figure 3-6 visualizes these legitimate state transitions as a Petri net[39].

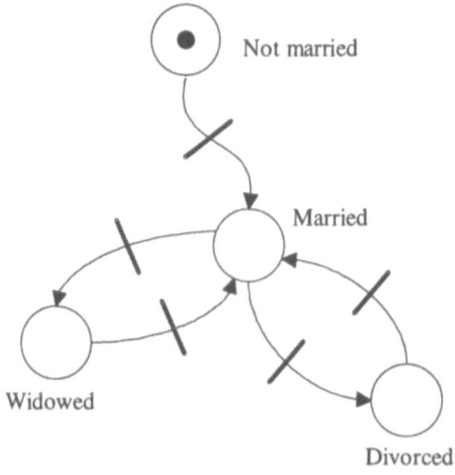

Figure 3-6:   Petri net specifying the allowed state changes for the property marital status

To automatically derive such a visualization, the specification of dynamic integrity constraints have to be based on a special syntax for the condition component. This encompasses a specification of the old state and a set of new states:

> *Syntax:*
> *ON*   ...
> *IF*   (property (<old value>) NOT TO (<set of new values>))
> *THEN* ...

The following business rules prescribe three of the permissible state changes of the property marital status:

---

[39] For the graphical representation of data behavior cf. Chapter 5.

Business rule 'Marital-status-1'
  ON   *(update marital status of person)*
  IF   **(marital status of person ('married') NOT TO ('divorced',**
         'widowed'))
  THEN   *reject operation;*
         raise event 'ILLEGAL STATE TRANSITION'

Business rule 'Marital-status-2'
  ON   *(update marital status of person)*
  IF   **(marital status of person ('divorced') NOT TO ('married'))**
  THEN   *reject operation;*
         raise event 'ILLEGAL STATE TRANSITION'

Business rule 'Marital-status-3'
  ON   *(update marital status of person)*
  IF   **(marital status of person ('widowed') NOT TO ('married'))**
  THEN   *reject operation;*
         raise event 'ILLEGAL STATE TRANSITION'

3.2.2.3   Action Types

For the action component of business rules several classifications have been proposed[40]. Similar to the event and condition part, they can be classified into *elementary* and *composite actions*.

- An *elementary* action consists of exactly one task which is not further refined.
- A *composite* action is a sequence of several tasks to be executed.

As mentioned above, actions may raise events; therefore, a further classification
of elementary actions according to the elementary event types is appropriate:

- *Data manipulation actions* encompass the creation, modification, retrieval,
  derivation, and deletion of data[41]. These data objects are not necessarily
  stored in a database but may also be stored outside the automated part of an
  IS (e.g. on a form). Insert and delete concern entire tuples and the other
  three may involve specific properties. Thus, one syntax for tuple operations,

---

[40] Mertens/Hofmann (1986); Hofmann (1987); Herbst/Knolmayer (1995), p. 152.
[41] In Ceri (1992), p. 453, another possible classification of data manipulation actions is
      described which distinguishes positive and negative actions. Positive actions may lead
      to the creation of new data objects and negative actions may lead to their deletion. In
      this rather simple approach only the creation and deletion of data objects can be
      classified unambiguously whereas modification and retrieval of data objects are not
      considered.

one for modifying property operations, and one for the retrieval has to be provided:

Syntax for tuple operations:
 ON ...
 IF ...
 THEN DELETE ¦ INSERT <object name>

Syntax for modifying property operations:
 ON ...
 IF ...
 THEN UPDATE ¦ DERIVE <property> := <expression>

Syntax for retrieval operations:
 ON ...
 IF ...
 THEN RETRIEVE <list of properties>

Example:
 ON (update stock of product)
 IF (stock of product < reorder-threshold)
 THEN insert order;
  update state of product := 'reordered'

- *User actions* encompass a task which may be related to a data object but does not imply one of the operations mentioned above. Similar to conditions in automation rules, user actions may contain any text; thus, a detailed syntax for their content would probably make the specification of certain actions impossible.

 Syntax:
  ON ...
  IF ...
  THEN <any text>

 Example:
  ON (update stock of product)
  IF (stock of product < reorder-threshold)
  THEN call supplier and reorder product

- *Message actions* consist of a message to a processor of the IS; this message may be issued to an application or a human actor. A business rule whose action can be classified as a message action is called an *alerter*[42]. The mes-

---

[42] Cf. e.g. Dayal (1988), p. 130.

sage can inform the target person on a situation and may trigger a specific action[43]. Thus, the syntax for message actions contains at least a message and sometimes also the recipient of the message:

Syntax:
    ON   ...
    IF   ...
    THEN   issue <message> [TO <processor>]

Example:
    ON   (update stock of product)
    IF   (stock of product < reorder-threshold)
    THEN   issue 'Please reorder' TO purchaser

Because executed actions may result in events, the classifications of elementary actions and events must be consistent, i.e., the interdependencies put together in Figure 3-7 have to be respected.

	may raise event of type		
	Data manipulation event	User event	Time point event
Data manipulation action Insert of order	Yes ⇨ Order inserted	No	No
User action Check quality of stock entry	No	Yes ⇨ Quality of stock entry checked	No
Message action „Stock is low, please reorder product"	No	Yes ⇨ Reorder necessary	No

Figure 3-7: Interdependencies between action and event types

The action component may be elementary or composite, i.e., a sequence of elementary actions. In the following the syntax for elementary actions is described.

---

[43] Cf. e.g. Mertens/Hofmann (1986), pp. 326.

3.3   Business Rules in their Organizational Environment

3.3.1   Origins of Business Rules

Business rules of an organization are artifacts of ethical, cultural, or legal norms or result from organization internal decisions. As discussed in Chapter 2, only a subset of all business rules is formalized in e.g. organizational guidelines; the others constitute the common knowledge of an organization and may only be explicitly specified for the development of an IS[44]. In order to comprehensively describe the requirements for an IS, all relevant business rules of the concerned organizational unit or business process have to be found and formalized. For the search for business rules, a large amount of potential sources, such as e.g., legislation, organigrams, or salary systems have to be analyzed. The relationship of business rules with origins supports the analyst in completely collecting these rules:

- Prior to collecting business rules, a list of potential origins can be assembled. This enables a search of these origins for relevant rules and makes the task more efficient.
- During gathering business rules, the analyst can check which origins have already been searched.
- The assignment of business rules to origin classes supports solving conflicts between business rules because the main origin classes may be ordered with respect to their validity.
- Independently from the method of their implementation, one major problem of rules in IS is their maintenance, i.e., they have to be kept consistent with the real world. When business rules in the real world change, all implementations of these rules in the IS have to be adapted. This may be supported by selecting all business rules which have been derived from a specific origin, e.g., an organizational guideline.

In Figure 3-8 the origins of business rules are classified on two levels[45]. On the first the origins are categorized with respect to their relation to the organization into external and internal origins. The second level encompasses a further classification of the two basic types.

---

[44] Cf. Knolmayer/Herbst (1993), p. 386.
[45] Cf. e.g. Lipeck (1989), p. 1.

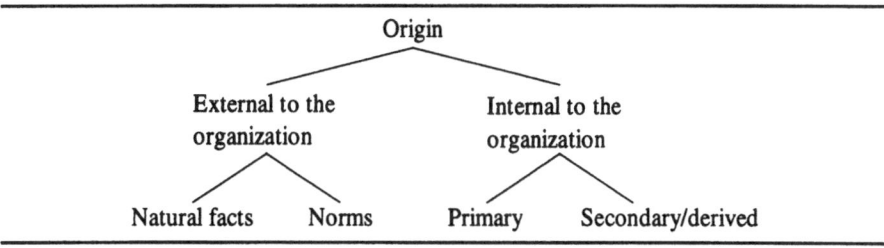

Figure 3-8: Origin classes

3.3.1.1 External Origins

The influence of an organization on the content of business rules resulting from *external origins* is very restricted; therefore, from the viewpoint of an organization, these rules can be considered as determinants, i.e., prescriptions for the business. External origins are further classified into *natural facts* and *norms*.

- *Natural facts:* Business rules based on natural facts never change over time; they are eternally fixed. The following two business rules contain facts which are based on this type of origin:

    ON   *(insert person) OR (update sex of person)*
    IF   *(sex NOT IN {male, female})*
    THEN *error*

    ON   *(insert person) OR (update date of birth of person)*
    IF   *(date of birth > current date)*
    THEN *error*

- *Norms:* External norms encompass among others legal, ethical, and cultural norms. Contrary to the examples specified above, business rules found in such origins may change in time. Furthermore, because norms depend on the values of a society, there may be differences between rule instances resulting e.g. from different environmental conscience or the employment legislation.

    ON   *(entry date of birth of a person)*
    IF   *(date of birth > current date - 14 years)   /* limit for child labor */*
    THEN *deny employment*

If an IS is used in several regions with e.g. different legislation, the criterion *origin* may help to find those rules whose content differs depending on the origin. For such a precise analysis, a further classification of norms may be done (e.g. legal norms in international and national law).

Business rules resulting from ethical and cultural norms differ from those resulting from legal norms with respect to their compulsory nature because an organization can decide whether to respect these norms or not[46]. The sanctions resulting from violating a norm depend on the situation of the society in which an organization acts. Therefore, the violation of certain environmental norms e.g. may be ignored during a recession.

3.3.1.2   Internal Origins

Business rules retrieved from *internal origins* may be valid on a long-, medium-, or short-term basis. Unlike rules from external origins, they may be modified within the limits set by validity duration and by potentially existing interdependencies. On the second level, internal origins are classified in *primary* and *secondary/derived origins*.

- *Primary origin*: An origin of a business rule is its primary origin if the specification of the rule has not been directly derived from another origin.
- *Secondary/derived origin*: This type contains business rules which are derived from another (possibly primary) origin.

This distinction between primary and derived origins cannot be generally done for an origin but depends on each specific business rule: From the origin 'organizational guideline' the analyst can retrieve rules for which this origin is a primary origin, as well as rules which have previously been derived from other origins.

Rules retrieved from primary origins have to be analyzed with respect to their validity and consistency. For business rules found in derived origins, additionally the consistency with the primary origin has to be checked. Therefore, in order to reduce the tasks of verifying and validating the specifications, the analyst should strive for the identification of primary origins for all business rules.

3.3.1.3   Conflict Situations

The large amount of business rules in an organization and their interdependencies can lead to inconsistencies. In the following, three conflict situations are discussed:

- *Rules of two different primary origins*: Their inconsistency may result from the intransparency and non-handled complexity of rules in an IS. The missing knowledge about rules specified in other origins may lead to the specification of inconsistent 'duplicates'.

---

[46] In Rückle/Terhart (1986), the observance of legal norms is discussed from the viewpoint of decision theory.

Legal norm:
An employee has to be at least 14 years old.
ON   *(update date of birth of person)*
IF   *(date of birth > current date - 14 years)*
THEN   *deny employment*

Internally formalized business rule:
An employee has to be at least 12 years old.
ON   *(update date of birth of person)*
IF   *(date of birth > current date - 12 years)*
THEN   *deny employment*

- *Rules of a primary and a derived origin*: Inconsistencies between a rule in a primary and a derived origin may result from a wrong derivation. The following example is based on a statement found in a guideline which is considered as the primary origin for the business rule:

  Primary origin:
  A product is to be reordered when the stock falls below a reorder-threshold.

  Business rule in the purchasing department:
  ON   *(update stock of product)*
  IF   *(new stock ≤ 0)*
  THEN   *reorder product*

- *Rules of two derived origins*: Inconsistencies between 'copies' of a business rule derived from the same primary origin may result from an different interpretation of an ambiguous rule in the primary origin. Again the reorder rule is used as an example:

  Primary origin:
  A product is to be reordered when the stock falls below a threshold.

  Business rules in the purchasing department:
  *The product is **only** reordered when the stock is below the threshold; earlier reordering is explicitly forbidden.*
  ON   *(update stock of product)*
  IF   *(new stock < reorder-threshold)*
  THEN   *reorder product*

  ON   *(reorder product)*
  IF   *NOT (new stock < reorder-threshold)*
  THEN   *reject reordering*

Business rule in the sales department:

> The product is **at last** reordered when the stock is equal or below the
> threshold; earlier reordering is possible.
> ON   update stock of a product
> IF   (new stock ≤ reorder-threshold)
> THEN   reorder product

The examples given above already indicate that the origin types can be ordered
with respect to the validity of rules (cf. Figure 3-9). However, even in applying
this order, one always has to check whether the dominant rule is really valid, i.e.,
consistent with the real world.

		Business rule$_2$ originates from a			
		Natural fact	Norm	Primary origin	Derived origin
Business	Natural fact	?	Natural fact	Natural fact	Natural fact
rule$_1$	Norm		?	Norm	Norm
originates	Primary origin			?	Prim. orig.
from a	Derived origin				?

Figure 3-9:   Validity of conflicting business rules depending on their origin types

3.3.1.4   Origin Owner

In the task of defining or validating business rules[47], the analyst has to know who
'owns' an origin and is therefore responsible for the content of all business rules
retrieved from that origin. The assignment of origins to organizational units and
the analysis of all rule components allows the analyst to distinguish two types of
business rules:

- *Single owner rule*: All rule components are found in origins which are
  owned by a single organizational unit.
    ON   (customer calls)   /* Product development department */
    IF   (exist support contract for customer) /* Prod. develop. dep.   */
    THEN   provide hot line support   /* Prod. develop. dep.   */

- *Multiple owner rule*: At least two components have to be specified by different organization units; a rule of this type represents a *responsibility interface* of two or more organizational units.
    ON   (update stock of product)   /* Storage department   */
    IF   (stock < reorder-threshold)   /* Storage department   */
    THEN   reorder the product   /* Purchasing department   */

---

[47]   Cf. Section 4.2.

3.3.2   Processors of Business Rules

3.3.2.1   Processor Types

Each business rule of an IS may be assigned to processors[48] which execute the rule components either manually or automatically. In the discussion of work flow management the notion of *processing entities* is used as a synonym[49]. *Manual* processors are human actors like e.g. a stock keeper. *Automated* processors are either software components in case of data processing or machines in case of automated manufacturing. If the assignment of a process to manual and automated processors is not made visible (cf. Figure 3-10), a process can not be described sufficiently. Thus, the distribution of business rule components on manual or automated processors makes the man-machine interfaces transparent.

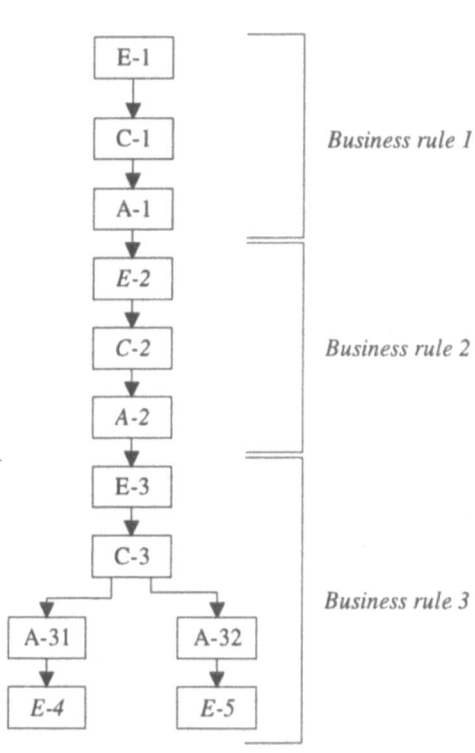

Figure 3-10: Process without assignment of business rules to processor types

---

[48] The term *processor* is not used in its technical meaning as a part of the computer hardware.

[49] Cf. Rusinkiewicz/Sheth (1995), p. 592.

In the following section, the interdependencies between business rules and processor types is discussed in three subsections. The first and second discuss the assignment of a single business rule to processor types and to specific processors, respectively. Afterwards, the discussion is enlarged on the distribution of an entire process consisting of business rules.

3.3.2.2   View on Single Business Rules

$2^4=16$ alternatives exist for the assignment of the four basic components of a business rule to either the automated (a) or the manual (m) subsystem. Because of the exclusive execution of the two action components, the 16 types of Figure 3-11 can be reduced to eight. The clustering of the 16 types is done with respect to the maximum number of interfaces between the components of a business rule:

- Types 2, 3, and 4 are classified as *aam*
- Types 5, 6, and 7 are classified as *ama*
- Types 10, 11, and 12 are classified as *mam*
- Types 13, 14, and 15 are classified as *mma*

Original type #	Final type	E	C	A	A	Max. number of interfaces	Number of automated components
1	aaa	a	a	a	a	0	4
2		a	a	a	*m*	1	3
3	aam	a	a	*m*	a	1	3
4		a	a	*m*	*m*	1	2
5		a	*m*	a	a	2	3
6	ama	a	*m*	a	*m*	2	2
7		a	*m*	*m*	a	2	2
8	amm	a	*m*	*m*	*m*	1	1
9	maa	*m*	a	a	a	1	3
10		*m*	a	a	*m*	2	2
11	mam	*m*	a	*m*	a	2	2
12		*m*	a	*m*	*m*	2	1
13		*m*	*m*	a	a	1	2
14	mma	*m*	*m*	a	*m*	1	1
15		*m*	*m*	*m*	a	1	1
16	mmm	*m*	*m*	*m*	*m*	0	0

Figure 3-11: Assignment of rule components to automated or manual processors

The assignment to manual or automated processing is evaluated separately for each of the three basic rule components:

- *Manual or automated event detection?*
  The detection of all events relevant in business rules requires the monitoring of a huge amount of situations in the real world, the applications, and the database. Therefore, it can be expected that a manual event detection leads to a rather insecure rule execution because the human actor himself has to monitor all situations and has to recognize each relevant occurrence of a possibly even composite event.
- *Manual or automated condition evaluation?*
  The correct evaluation of the condition is very important for the action to be executed. However, there may exist fuzzy conditions which cannot be specified exactly. The evaluation of a condition, such as the one specified in the following example, may be supported by a decision support system but will probably never be completely done automatically.

  > ON   *(insert customer)*
  > IF   *(customer is credit-worthy)*
  > THEN   *set credit-limit := unlimited*

- *Manual or automated action execution?*
  The action execution can also be assigned to a human actor or an automated component of the IS. The problem with a manual execution is again the missing security of processing. If the action itself does not lead to changes in e.g. the database, it is impossible for the automated system to check whether the action has been (correctly) executed. However, if the action changes the state of the database, another rule for checking the action execution could be formalized which could remind the manual processor:

  > ON   *(update stock of product)*
  > IF   *(new stock < reorder-threshold)*
  > THEN   *message 'Product runs out of stock. Please reorder'*
  >   *raise event 'REORDER-NECESSARY'*

  > ON   *2 Days after (REORDER-NECESSARY)*
  > IF   *(purchase order not registered)*
  > THEN   *message 'Product has run out of stock. Please reorder*
  >   *immediately'*

Application of the Classification

The assignment of business rules to the automated or manual subsystem can be done from three viewpoints:

- Where are the components *currently* processed?
  ⇨ *Current system description*

- Where can the components *potentially* be processed?
  ⇨ *Definition of alternatives*
- How are the business rule components *optimally* processed?
  ⇨ *Target system description*

The assignment of the target system is restricted by the available alternatives, i.e., if a component cannot be processed automatically its assignment to the automated subsystem of the target system is not allowed (e.g. an event like 'phone call of customer' can normally only be detected manually).

High-level business policies are normally transformed into low-level computational representations and are buried deep within the system program code[50]. To support the maintenance of an IS, the examination of the *current assignments* helps to find the (possibly redundant) implementations of a specific business rule: the modification of e.g. an organizational guideline leads to the determination of business rules originating from this document. Afterwards, the actual implementation can be obtained by regarding current assignments of rule components to processors of the automated subsystem, i.e., specific programs.

The three viewpoints are exemplified with the following business rule:

ON   *(update stock of product)*
IF   *(new stock < reorder-threshold)*
THEN   *reorder product*

- *Current system:* The assignment of the current system results from examining its current processing. The rule could e.g. currently be assigned to type *aam*, i.e., the event and condition are processed automatically and the product is reordered manually.
- *Definition of alternatives:* Each of the three components can be processed either in the manual or the automated subsystem, i.e., no restrictions exist for the specification of the target system:
  - *Event:* The reduction of the stock can be detected and signaled either in the manual subsystem (e.g. by the stock keeper) or automated by a function (e.g. a database trigger) which controls the stock.
  - *Condition:* Depending on whether the information needed for the evaluation of a condition is stored in the automated system, the condition can be checked automatically or manually.
  - *Action:* For the action both assignments are also possible; therefore a product may be reordered manually or automatically (e.g. by using electronic data interchange).

---

[50] Poo (1992), p. 95.

- *Target system*: From an global, organizational view on the target system, all business rules have to be considered; however, for the implementation of applications only those rules of the target system which are not assigned to type *mmm* have to be considered. The assignment of components to either a manual or an automated execution has to take several aspects into account[51]:

  - *Flexibility*: Rule components implemented in a computer system are processed very strictly; this may prevent adequate handling of exceptions[52].

  - *Consistency of rule processing*: Assuming a correct implementation of a rule component in the automated subsystem, this processing is standardized and always consistent. An execution by a manual processor may differ and is not always controllable by the automated system. Especially the execution on schedule is hard to control.

  - *Data*: How difficult is it to store all data that is needed for a fully automated processing of a rule component? Because of the potential difference between the assignments for the current and the target system, different requirements for data stores may result.

  - *Complexity*: Is it possible to implement the logic of a component with acceptable expenses? Examples for critical algorithms are nonfeasible ones like sequence scheduling or unstructured conditions like the evaluation whether a potential debtor is credit-worthy.

  - *Number of interfaces*: Interfaces from *manual* to *automated* processors may be implemented by setting specific flags or by manually invoking functions of an application. The communication from the automated subsystem to a manual processor may involve signals, e.g., screen messages or electronic mail. Each of these interfaces between automated and manual subsystem incorporates potential errors because not only the execution of a rule component has to be correct, but also the exchange of the accompanying information. Therefore, the amount of interfaces necessary for processing a business rule should be minimized and one should strive for a homogenous rule processing. From this point of view, types *aaa* and *mmm* are optimal and types *ama* and *mam* should be avoided.

  - *Desirability*: Besides technical aspects, one has to determine whether the control on processing a specific rule component should be left to a computer system (e.g. with respect to ethical reasons).

---

[51] Cf. Herbst/Knolmayer (1995).
[52] Cf. e.g. Strong/Miller (1995), pp. 207.

In this context of planning the assignment of rule components to the manual and automated subsystem, ideas of 'Business Systems Planning'[53] may be applied. This approach tries to assign data classes and business processes to subsystems in such a way that interdependencies between different subsystems are minimized, subsequently minimizing the amount of interfaces. Similarly, business rules may be clustered to subsystems with the goal to minimize the interfaces between them. The boundaries resulting from such an assignment will be congruent with business processes rather than e.g. with a divisional or functional organizational hierarchy.

3.3.2.3   Assignment of Business Rules to Specific Processors

Additionally to processor types, the assignment of business rules to processors themselves may be analyzed, leading to *single* and *multiple processor* rules:

- *Single processor rule*: All rule components are executed by a single processor, i.e., there are no interruptions in the work flow.

ON	*(customer calls)*	/* Hot line supporter */
IF	*(exist support-contract for customer)*	/* Hot line supporter */
THEN	*provide hot line support*	/* Hot line supporter */

    In this rule, all components are processed by a single processor who or which receives the phone call, checks the existence of a support contract, and supports the customer. Regarding the examples in Section 3.2, this example shows that the responsibility for the content and for the processing of the rule may differ.

- *Multiple processor rule*: At least two components are executed by different processors. This leads to an interruption and thus e.g. information or material have to be passed to the processor who or which continues the execution of the business rule.

ON	*(update stock of product)*	/* Sales clerk */
IF	*(stock < reorder-threshold)*	/* Store's supervisor */
THEN	*reorder the product*	/* Purchaser */

    This business rule is triggered by an event detected by a sales clerk in the sales department, the condition is checked by the store's supervisor in the storage department, and the resulting action is performed by the purchaser in the purchasing department.

---

[53] For methods of Business Systems Planning cf. e.g. IBM (1980), pp. 48; Knolmayer/Spahni (1993), pp. 99.

3.3.2.4   View on Processes Specified Using Business Rules

In the preceding subsection, the assignment of single business rules to manual or automated processors has been discussed. This view is now extended to the whole business process described by using business rules. The distribution of a (sub-) process on the manual or automated subsystem of an IS results from the assignment of the constituting business rules to manual or automated processors (cf. Figure 3-12). This more global view is especially relevant for a process orientation as attempted in BPR projects[54]. In this context, the analysis of the interfaces resulting from the process distribution may help to optimize the process execution.

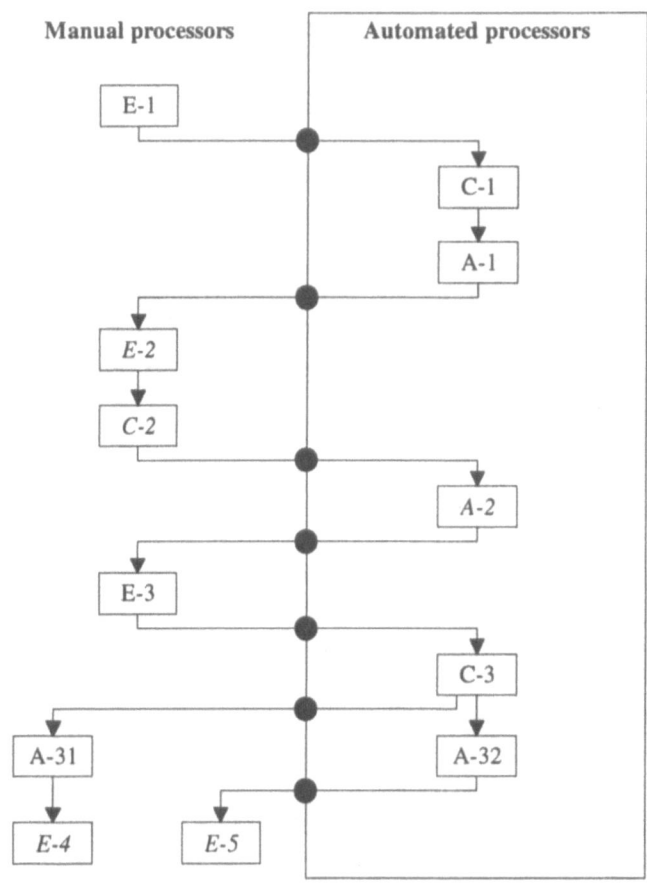

Figure 3-12: Distribution of a process on manual and automated processors

---

[54]   Cf. e.g. Hammer/Champy (1993), pp. 35.

Regarding this distribution two process classes can be distinguished:

- *Single processor process*: All rule components of the process are processes by a single processor, i.e., the work flow is not interrupted.
- *Multiple processor process*: At least two components of the process are executed by different processors which leads to an interruption of the work flow.

Such an analysis of processes supports the exact specification of processor-interfaces within a business process. Thus, in order to achieve a continuous work flow, the number of interfaces involved in the execution of business rules constituting a process should be minimized.

3.3.3 Assignment of Business Rules to Organizational Units

The use of an IS is often not restricted to a single organizational unit. Therefore, many business rules, and thus also processes, may result in interfaces between organizational units. In order to relate business rules and processes to the organizational structure, manual and automated processors are assigned to organizational units (cf. Figure 3-13), thereby deriving the (indirect) assignment of business rules and thus entire processes to organizational units (cf. Figure 3-14).

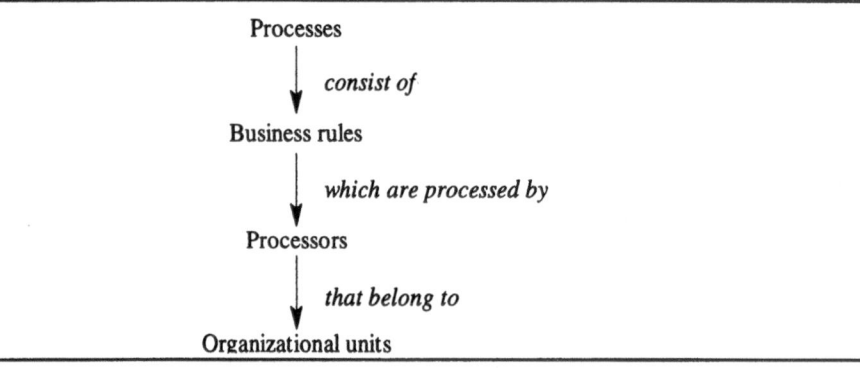

Figure 3-13: Indirect relationship between processes and organizational units

The distribution of business rules of a process on organizational units leads to two different process types:

- *Intra-unit process*: Processes which are executed within a single organizational unit are easier to execute and to maintain because no interfaces between organizational units exist and thus, no additional communication between units is necessary. Therefore, it can be argued that process execution within an organizational unit is optimal and may result in organizational structures defined with respect to process boundaries.

- *Inter-unit process*: Processes which overlap the boundaries of organizational units may lead to two questions: (1) are the processors assigned to the correct organizational units and (2) are the boundaries of the organizational units defined correctly?

As an example, the execution of a single business rule may involve human actors of three different departments:

ON	*(update stock of product)*	*/* Sales clerk	*/*
IF	*(stock < reorder-threshold)*	*/* Stock keeper	*/*
THEN	*reorder the product*	*/* Purchaser	*/*

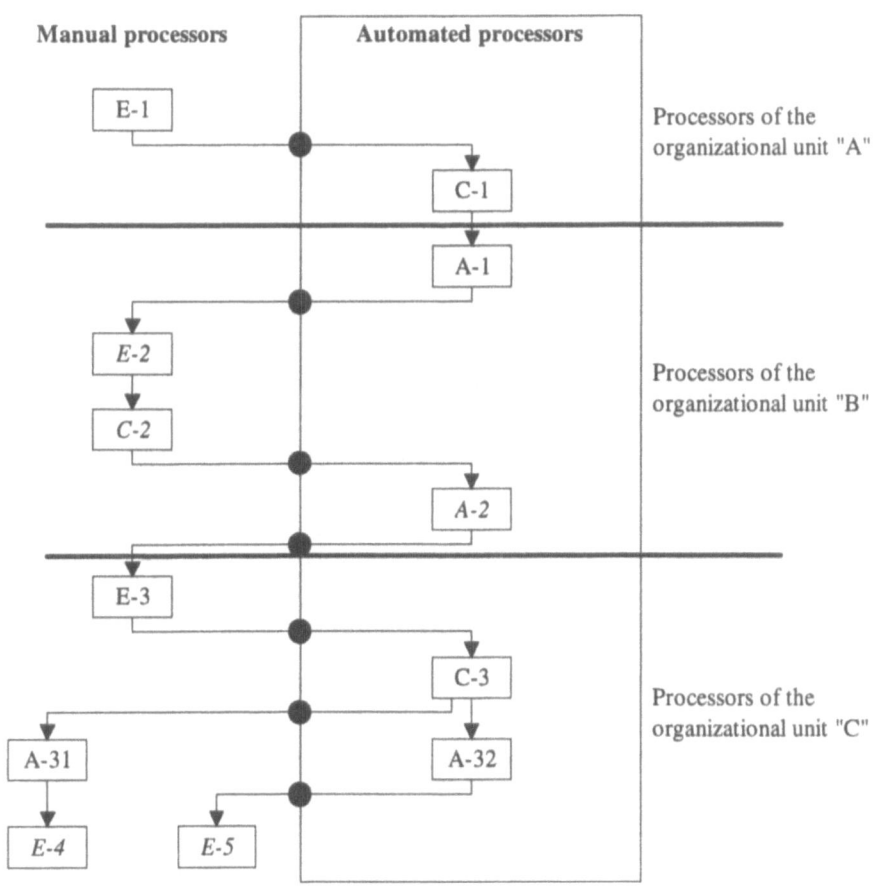

Figure 3-14: Relation of a process to organizational units

3.4   Summary

This chapter has dealt with four aspects: (1) the case studies and examples of business rules, (2) the classification of integrity constraints and business rules, (3) the syntax for the specification of business rules, and (4) the relationships of business rules with other organizational facts.

The case studies that were carried out revealed information about the quantities and types of business rules in real IS. Solving the quantity problem can be seen as one of the main requirements for the *BROCOM* approach. One possibility to reduce this complexity is to classify business rules. The classification system presented first divided business rules into *automation rules* and *integrity rules*. Because integrity rules encompass the semantic of integrity constraints, selected criteria for their classification were taken into consideration. Additionally, different types of events, conditions, and actions were introduced and used as a basis for the definition of a syntax for each of the component types.

A second promising alternative for complexity reduction is to link business rules with closely related facts which encompass origins, processors, and organizational units. *Origins* are either organization external (e.g. a part of the legislation) or internal (e.g. an organizational handbook). The knowledge about the relationship between business rules and origins facilitates finding the real world equivalent of a business rule and thus enables the analyst to validate this rule. Furthermore, the rule set can be divided into rules that are fixed (organization external) and rules that can be influenced (organization internal). *Processors* are either manual (i.e. human actors) or automated components (e.g. application programs) of the IS which execute specific parts of business rules. The assignment of business rules to processors makes the distribution of a process on different processors and processor types manageable. *Organizational units* are responsible for origins and contain processors. Thus the interdependencies between business rules and the organizational structure become visible.

This chapter has focused on the concept of business rules and has shown the general ideas of the *BROCOM* approach. Based on this, the next chapter encompasses the presentation and a first small application of the *BROCOM* approach.

4 *BROCOM*: A BUSINESS RULE-ORIENTED CONCEPTUAL MODELING APPROACH

In this chapter, *BROCOM*, an approach for a business rule-oriented conceptual modeling is introduced. The *BROCOM* approach emphasizes the use of business rules which are used for the specification of all dynamic properties relevant to the universe of discourse, i.e., of processes *and* integrity constraints. This focus on a single modeling construct reduces the complexity of the conceptual model and does not force the analyst to artificially separate facts that belong together. The approach is presented in two main sections which encompass the *meta model* and the *modeling steps*.

The *meta model* presented is based on the discussion of business rules and related facts as presented in Chapter 3. The introduction of the business rule meta model is complemented by a comparison with meta models of other approaches for conceptual modeling. The steps of *BROCOM* are described in detail by using business rules themselves as specification constructs. To illustrate the practical application of the approach, each step is directly applied to the example of the case study DOM-2.

4.1 The Meta Model of the *BROCOM* Approach

4.1.1 Introduction

The systematic development of IS is often supported by tools for computer aided software engineering (CASE) which encompass support for different techniques. The meta data that is manipulated by these tools is usually stored in a special database, the repository. Repository systems can therefore be seen as specialized DBMS with some special services to administer meta data; terms related to *repository* are *data dictionary* or *directory*[1].

For the definition of an information resource dictionary system[2] (IRDS), a standard has been established by committees of ANSI[3] and ISO[4]. The ANSI standard

---

[1] Cf. Myrach (1995), pp. 1.
[2] The term dictionary is often used as a synonym for repository.

encompasses seven modules; however, to fulfill the requirements of the standard, only the functionality of the core module has to be implemented[5]. Common to both standard approaches is the following 4-layer-architecture:

1. *Dictionary definition schema layer*: Definition of the basis constructs which are available for describing information; the content of this layer is defined by the standard.
2. *Dictionary definition layer*: Information about the meta structure of facts which are stored in a data dictionary (e.g. the meta entity type 'Entity type').
3. *Dictionary layer*: Information about the data of the real world (e.g. the entity type 'Customer').
4. *Application layer*: Facts of the real world (e.g. entities as the customer 'Miller').

Meta models as those discussed in this chapter are assigned to the second layer, i.e., they describe the meta structure of facts from the universe of discourse.

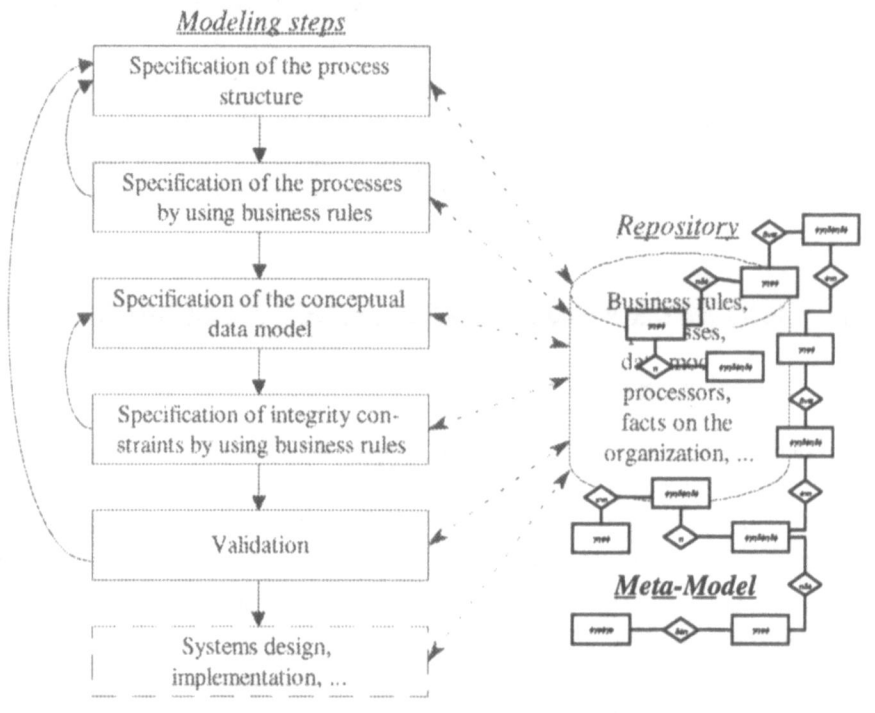

Figure 4-1:   Framework of BROCOM: Meta model

3   Cf. ANSI (1989).
4   Cf. ISO (1989); Spurr (1989).

For the conceptual specification of this second IRDS layer, the boundaries of the universe of discourse and the type of relevant facts need to be defined. In the context of IS development, the universe of discourse is mainly determined (1) by the stage of modeling (e.g., systems analysis, systems design, or implementation), and (2) by the type of IS that will be developed.

The *BROCOM* meta model[6] described in this chapter is to be used for the conceptual modeling of IS in the stage of systems analysis. As described in the preceding chapters, the modeling of business rules, processes, and data structures are the main areas of interest which additionally ought to be linked to organizational facts. The structure of business rules has already been defined above; therefore, we now need to look at the facts which are closely related to them:

- *Data model*: Which conceptual data model should be chosen?
- *Organization*: Which components within the organization are linked to business rules, are relevant for the IS development, and therefore have to be considered?
- *IS*: Which IS components are related to business rules and have to be included in a conceptual model?

The main goal of the meta model described in this Chapter is to provide a structure which allows for the storage of all meta data needed for a conceptual modeling of business rules. Furthermore, the meta model should allow for derivation of different graphical or verbal representations of views on the meta data. These views should encompass a process view, a data structure view, a data usage view, and a data behavior view[7].

The meta model is presented in two steps: First, the 'core' of the meta model is discussed. It includes the main constructs *Business rule, Process*, and *Data model component*. The other submodels, i.e., *Origins, Processor*, and *Organizational unit* are discussed as the 'environment' of business rules[8]. For the specification of the cardinalities, the meta model employs a *min,max* notation. As an example, Figure 4-2 represents the facts that a customer is related to zero, one, or several bills and that a bill is always assigned to exactly one customer. The roles of the relationships have to be read from the upper to the lower entity-type or from the left to the right one, respectively.

---

6   Cf. Herbst (1996).
7   Selected alternatives for the representation of these views are the subject of Chapter 5.
8   A related model is given in Joosten (1994), p. 238; however, this model does not incorporate a condition components and neglects organizational units.

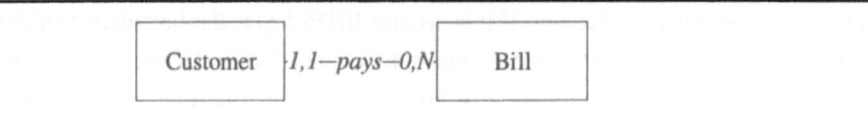

Figure 4-2: Example of the cardinality notation

4.1.2   The Core of the Meta Model

4.1.2.1   Submodel '*Business Rule*'

Business rules structured as ECA^2 rules consist of the three basic component types event, condition, and action. This is reflected by the meta entity type *Business rule* which is related by the *is aggregated of* relationship type to the three meta entity types *Event, Condition,* and *Action* (cf. Figure 4-3).

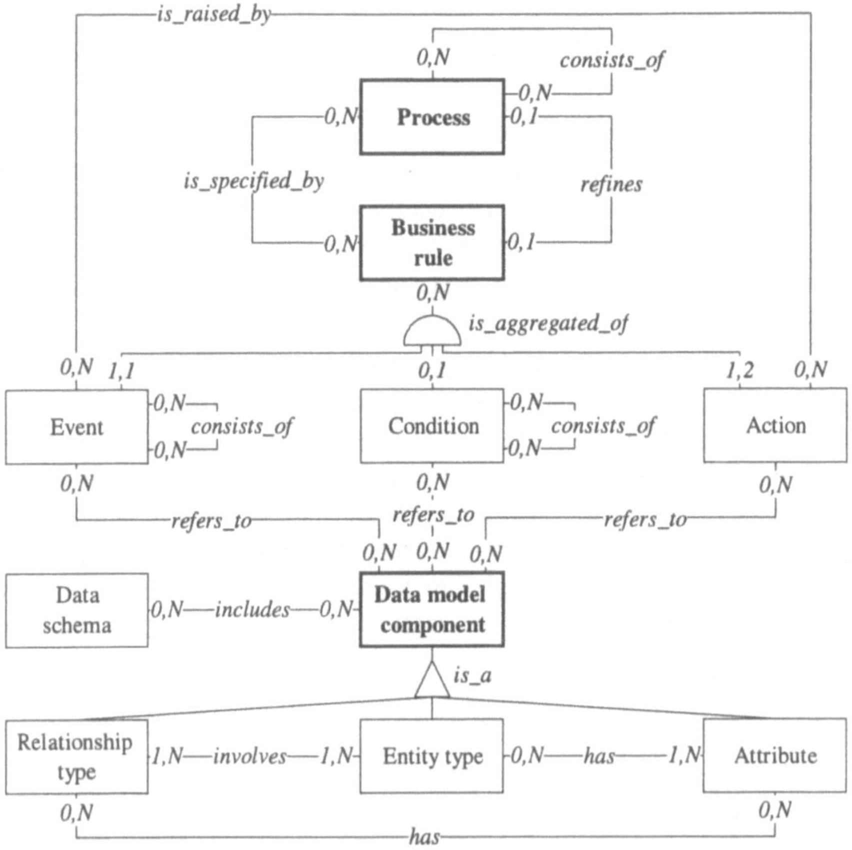

Figure 4-3: The core of the meta model

Every business rule has exactly one event, at most one condition, and one or two actions (then/else). As discussed in Chapter 3, events and conditions may be composite and therefore have recursive m:n relationship types. Actions of business rules may raise events. In the model this is represented by the relationship *is raised by* between the meta entity types *Action* and *Event*.

4.1.2.2   Submodel '*Process*'

Processes are a main element of organizations and of organizational approaches like business process reengineering. Therefore, in order to build an adequate model of the universe of discourse, they have to be emphasized in conceptual modeling. As already sketched in the examples given in Chapter 3 and as will be further shown in the following chapters, processes can be specified using business rules. In the context of this meta model only the behavior of a business process is considered. Additional properties like process goals, values, and process owners are not further discussed[9]. In the meta model two relationship types exist between processes and business rules:

1. Each process *is specified by* a finite set of business rules and
2. a process may *refine* a business rule[10].

Process Specification

Business rules formalized according to the ECA^2 structure may constitute chains of rules. These chains are constituted dynamically by events that are raised in actions and may trigger succeeding business rules. The result of the process specification is a finite set of business rules which *specify* a process. Because there may exist business rules which are not explicit elements of a process (e.g. business rules encompassing integrity constraints), the minimum cardinality of the relationship type from *Business rule* to *Process* is zero. On the other hand, there may exist processes which are only described by additional properties, e.g., process names and owners and have not yet been specified by using business rules.

Process Refinement

The case studies presented in Chapter 3 revealed the huge amount of business rules in a real IS as a major problem. This results in the need for reducing the complexity of rule and process specification. A common approach to reduce such complexity is to distinguish different abstraction levels, i.e., the refinement of facts on lower levels[11]. For such a complexity reduction in data flow diagrams, De Marco[12]

---

[9]   Cf. e.g. Hammer/Champy (1993); Davenport (1993); Jacobson et al. (1995). For an evaluation of meta models which focus on BPR cf. Hess/Bracht (1995).

[10]   An approach to refine behavior diagrams has been introduced by Schrefl (1990).In Theisges/Denk (1991) hierarchical state transition diagrams are discussed.

[11]   For refinement of Petri nets cf. Rosenstengel/Winand (1982); Schrefl (1990).

De Marco[12] specifies a balancing rule for leveled data flow diagrams: the parent
and the child diagram have to be *balanced,* i.e., „data flows into and out of the
parent bubble are equivalent to data flows into and out of the child diagram". An
adaptation of this rule for the refinement of business rules leads to the requirement that on both levels the triggering event and the raised events have to be
identical[13].

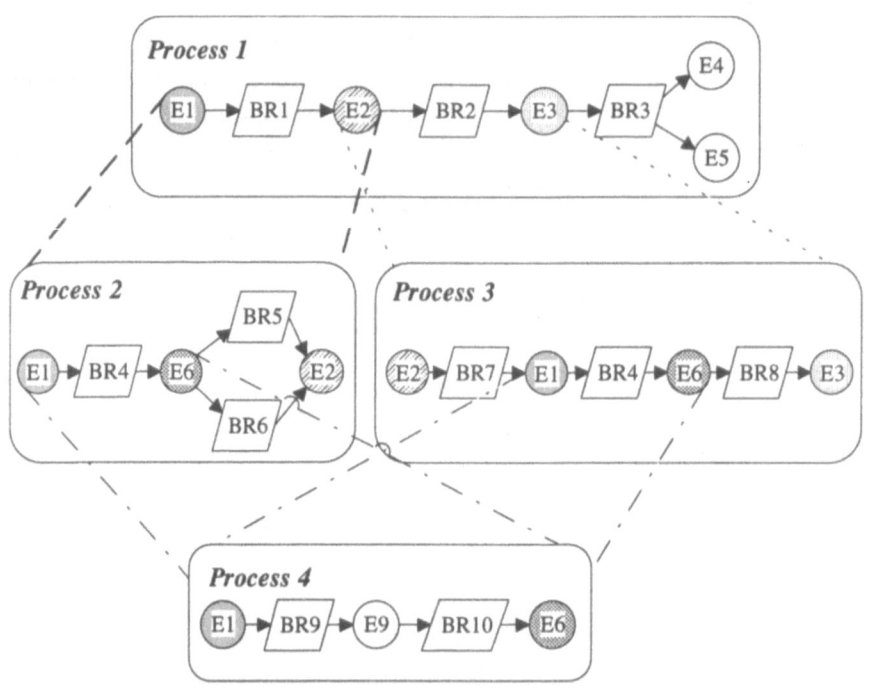

Figure 4-4:   Refinement of business rules and processes

Each process may have subprocesses and each of them may be specified by a finite set of business rules. Figure 4-4 depicts a simple event rule net, in which circles represent the events and parallelograms the condition and action components
of business rules[14]. In the diagram, a top-level *process 1* consisting of [BR1],
[BR2], and [BR3] is specified; [BR1] and [BR2] are refined by *processes 2* and *3*
on a lower level. This leveled approach allows for a modular process specification
and a reuse of business rules. Reused business rules are always triggered by the
same event(s) and always raise identical events; thus they may be regarded as
modules which can be included in different process specifications. However, de-

---

12 Cf. De Marco (1978).
13 Cf. Herbst (1995).
14 For details on the event rule net cf. Tanaka (1992).

act on raised events. [BR4] which is part of *process 2* and *3* leads to the event [E6] which triggers [BR5] and [BR6] in *process 2* and [BR8] in *process 3*, respectively. If such a reused business rule is further refined, the resulting lower-level process is automatically a subprocess of *all* processes which encompass the reused rule. As an example, [BR4] is additionally refined by *process 4* which therefore is a subprocess of *processes 2 and 3*.

The relationships between processes and business rules can be analyzed from two different viewpoints, i.e., from business rules to process and vice versa.

- Process ⇨ business rule
  - Which business rules are contained in a specific process?
  Example:
     The process 'Damage reception' is specified by the business rules of Chapter 3.

  - Which business rule is refined by a specific process ?
  Example:
     The process 'Damage reception' is a refinement of the business rule 'DAMAGE-RECEPTION' which is part of the main process of DOM-2.

- Business rule ⇨ process
  - Which processes encompass (components of) a specific business rule?
  Example:
     Within the damage treatment an event may be raised which indicates that the terms of a policy have to be adapted. This event is raised in the subprocess 'Damage treatment' and triggers a business rule in the process 'Policy adaptation' which is executed independently from the damage administration.

  - Which processes are a refinement of a specific business rule?
  Example:
     The business rule 'PERSON-CONTACTS-US' is not refined by a process.

4.1.2.3   Submodel '*Data Model Components*'

The submodel '*Data model components*' encompasses the meta entity types for a conceptual data model. The most popular data model for conceptual modeling is currently the Entity Relationship Model (ERM); therefore, the meta entity types *Entity type*, *Relationship type*, and *Attribute* are incorporated in this submodel (cf. Figure 4-3).[15]

---

[15]   In Chapter 5, the adaptation of the meta model to express a NIAM is presented.

4.1.2.3   Submodel '*Data Model Components*'

The submodel '*Data model components*' encompasses the meta entity types for a conceptual data model. The most popular data model for conceptual modeling is currently the Entity Relationship Model (ERM); therefore, the meta entity types *Entity type*, *Relationship type*, and *Attribute* are incorporated in this submodel (cf. Figure 4-3).[15]

The reference of business rules to components of the data model is either a read or a write operation. The allowed semantics of *refers to* relationships between components of business rules and data model constructs is put together in Figure 4-5.

Relationship from	Retrieval	Modification
Event ⇨ Data model component	No	No
Condition ⇨ Data model component	Yes	No
Action ⇨ Data model component	Yes	Yes

Figure 4-5: Relationships between ECA^2 rules and data model components

The submodels '*Business rule*' and '*Data model component*' enable a view on the meta model focusing on the relationship between modeling constructs and business rules; it can be regarded from the following points of view:

1. Business rule ⇨ modeling construct: What sort of impact has a specific business rule on modeling constructs?
2. Modeling construct ⇨ business rule: Which business rules use a specific modeling construct?

This analysis can be extended to processes specified by business rules, thus resulting in the (indirect) relationship between processes and components of the data model.

4.1.3   The Environment of Business Rules

In Chapter 3 facts closely related to business rules were discussed which ought to be included in the *BROCOM* meta model. Figure 4-6 depicts the relationships between already introduced and the following additional submodels:

- '*Origin*': Each business rule has at least one origin.
- '*Processor*': The execution of business rules in organizations is done by automated or manual processors, e.g., 'sales representative'.

---

[15] In Chapter 5, the adaptation of the meta model to express a NIAM is presented.

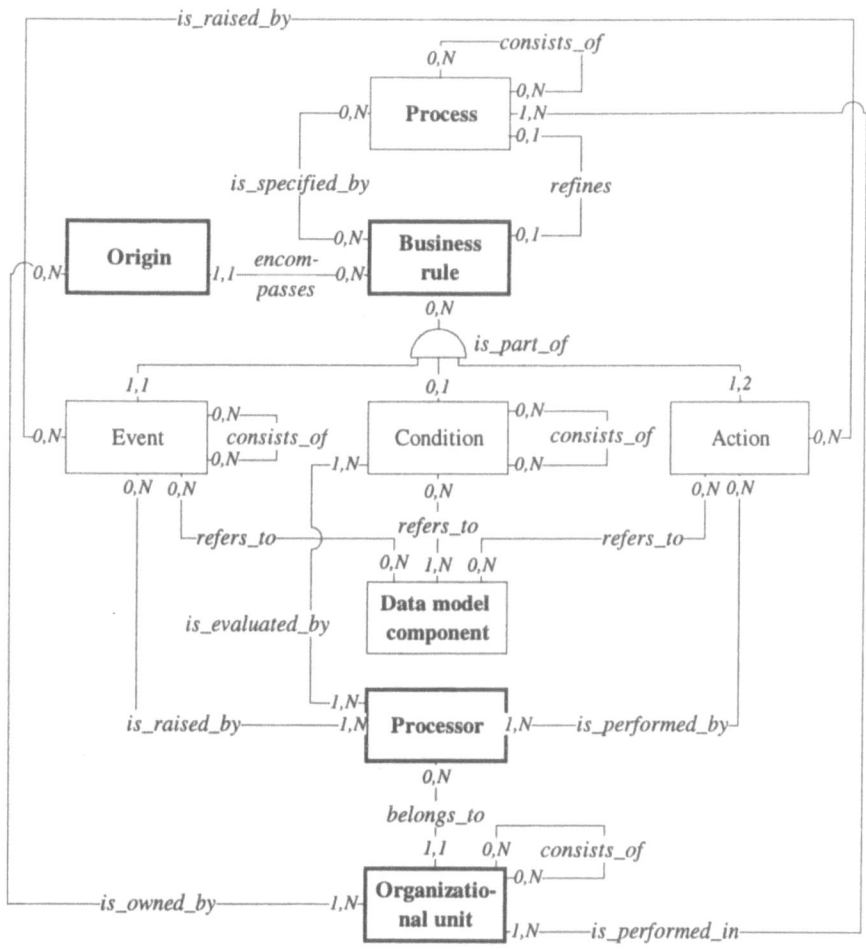

Figure 4-6:   Embedding business rules into their environment

4.1.3.1   Submodel '*Processor*'

The submodel '*Processor*' links rule components to specific processors who/which execute them, i.e., who/which detect the occurrence of events, evaluate conditions, and perform actions. Figure 4-7 depicts three processors types from which only actors manually execute rule components, whereas machines and software programs provide an automatic processing.

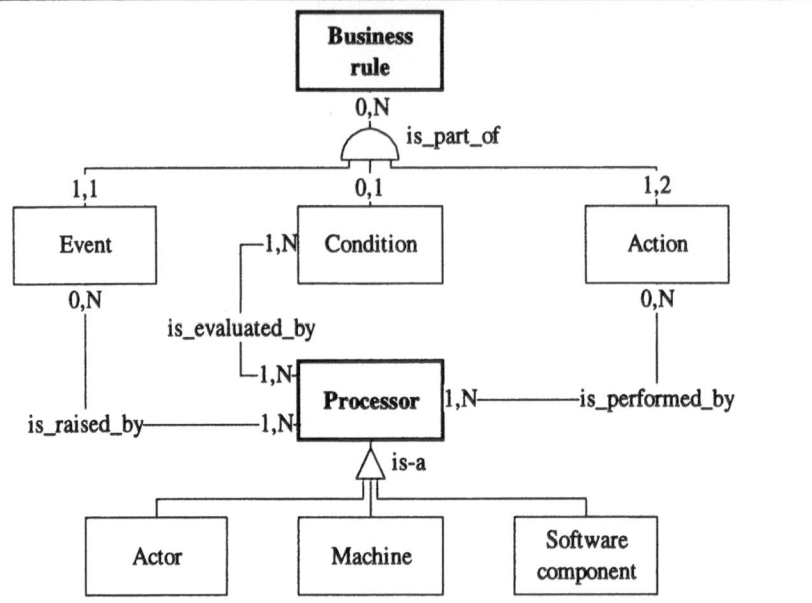

Figure 4-7:   Details of the submodel 'Processor'

The relationship between business rules and processors can be analyzed from two points of view:

- Processor ⇨ Business rule
  - Which events, conditions, or actions are executed by a specific processor?

  Example:
  > *A damage processing clerk is responsible for executing business rules 'REGISTER-PROV-DAMAGE', 'ACCEPT-PROV-DAMAGE', ...*

- Business rule ⇨ Processor
  - Which processor has to execute a specific business rule component?

  Example:
  > *Business rule 'REGISTER-PROV-DAMAGE' is executed by a damage processing clerk.*

4.1.3.2   Submodel '*Origin*'

Business rules may originate outside or inside an organization. The knowledge of the origin of a business rule also facilitates an analysis from both viewpoints:

- Origin ⇨ Business rule
  - Which business rules originate from a specific origin (e.g. in case of modifications of real world rules from this origin)?

Example:

From the origin 'policy third-party insurance' a specific subset of those business rules is derived, which checks whether a damage is covered by a policy.

- Business rule ⇨ Origin
  - Where does a specific business rule originate from (e.g. for checking the consistency between the rule and the real world)?

Example:

The business rule 'TIME-OUT-FOR-DAMAGE-FORM' is verbally defined on page 13 of the process description document 'DOM-process'.

4.1.3.3 Submodel '*Organizational Unit*'

In order to have a comprehensive model of the universe of discourse, the organizational structure has also to be considered. As discussed in Chapter 3, organizational units *own* origins and *encompass* processors which allows for the classification of rules and processes with respect to elements of the organizational structure. The analysis of the relationship between business rules and organizational units can again be done from different points of view:

- Organizational unit ⇨ Origins/processors
  - Which origins are owned by a specific organizational unit?

Example:

The 'Product development department' owns the terms of all policy contracts. It is thus responsible for most of the business rules collected in the case study DOM-2.

- Origin ⇨ Organizational unit
  - Which organizational units are responsible for the content of a specific business rule?

Example:

The terms of 'vehicle-insurance policies' are owned by the 'Product development department' which has to be involved e.g. to validate these rules.

- Processor ⇨ organizational unit
  - To which organizational units does a specific processor belong?

Example:

The actor 'damage-administration-clerk' belongs to the 'Damage processing department', i.e., all rules to be executed by this actor are processed within this department.

A well known approach for process modeling is part of ARIS[16], (Architecture for IS) which was developed at the Institute of Information Systems of the University of Saarbruecken. This approach provides a meta model which among other things contains a detailed submodel for storing information about organizational structures. This submodel may also be used as the organizational submodel within the *BROCOM* meta model.

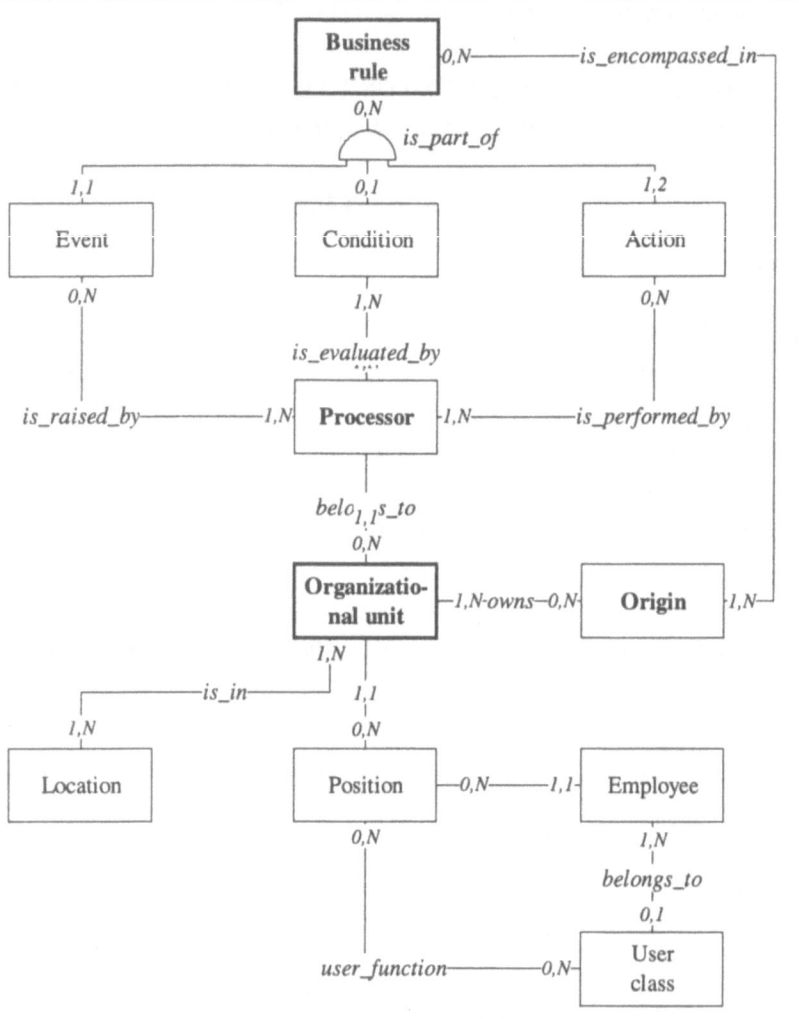

Figure 4-8:   Submodel 'Organizational unit' adapted from ARIS[17]

[16] Cf. Scheer (1991), pp. 90.
[17] Cf. Scheer (1991), pp. 92.

4.1.4   Selected Meta Models

The meta model for business rules presented above encompasses a focus on business rules structured as ECA^2 rules. Some components of the meta model, e.g., the meta entity types for the ERM, are also used in other meta models for the conceptual specification of IS. In order to compare the business rule meta model with existing approaches, four selected meta models are presented in the following which

- are intended to be used for conceptual modeling,
- focus on the development of business IS,
- are fully specified, and
- encompass the notion of events[18].

The discussion of the meta models is confined to the representation of the semantics of business rules and the related organizational facts.

In the representation of the meta models selected, cardinalities are given with the same precision as in the referred literature. Thus either the minimum and maximum or only the maximum cardinality is given. In the first case, the semantics of the notation is similar to the one used in the business rule meta model and in the second case, *1* stands for none or one and *n* stands for one or many.

4.1.4.1   Description of the Selected Meta Models

4.1.4.1.1   IFIP WG 8.1

In the last ten years, the IFIP working group 8.1[19] has been developing a framework for IS methodologies. The IFIP framework divides the IS life cycle in the four stages information systems planning, business analysis, systems design, and construction design[20]. In this framework the business rules approach is assigned to the stage *business analysis,* which is defined as the „examination and study of the existing state of affairs in a given business area of the enterprise"[21]. The IFIP WG 8.1 framework distinguishes a

---

[18] An additional framework in which *Brocom* can be regarded, is provided by the Euromethod described in Franckson (1994). In this framework, events are an explicit element of the business information view, the business process view, and the work practice view. Furthermore, business rules are used which govern the business process.

[19] The working group 8.1 is entitled 'Design and Evaluation of Information Systems' and is a subgroup of the technical committee 8, 'Information Systems'. The conferences of this working group took place in 1982, 1983, 1986, 1988, and 1994. They are well known under the acronym CRIS, standing for Comparative Review of Information Systems design methodologies.

[20] Cf. Olle et al. (1988), p. 13. The four main stages are extended on pp. 37 by eight additional stages, one of them prior to the four mentioned and seven after them.

[21] Olle et al. (1988), p. 15.

4.1.4.1   Description of the Selected Meta Models

4.1.4.1.1   IFIP WG 8.1

In the last ten years, the IFIP working group 8.1[19] has been developing a framework for IS methodologies. The IFIP framework divides the IS life cycle in the four stages information systems planning, business analysis, systems design, and construction design[20]. In this framework the business rules approach is assigned to the stage *business analysis,* which is defined as the „examination and study of the existing state of affairs in a given business area of the enterprise"[21]. The IFIP WG 8.1 framework distinguishes a

- data-oriented,
- process-oriented, and
- behavior-oriented

perspective. Furthermore, a cross-reference perspective is provided which includes the link between the three main views.

In order to have a comprehensive representation of the IFIP WG 8.1 meta model, Figure 4-9 puts together the meta entity types of all four perspectives provided by Olle et al.[22] Furthermore, to reduce the complexity, entity types representing m:n relationships are omitted and drawn as relationship types.

---

[19]   The working group 8.1 is entitled 'Design and Evaluation of Information Systems' and is a subgroup of the technical committee 8, 'Information Systems'. The conferences of this working group took place in 1982, 1983, 1986, 1988, and 1994. They are well known under the acronym CRIS, standing for Comparative Review of Information Systems design methodologies.

[20]   Cf. Olle et al. (1988), p. 13. The four main stages are extended on pp. 37 by eight additional stages, one of them prior to the four mentioned and seven after them.

[21]   Olle et al. (1988), p. 15.

[22]   Cf. Olle et al. (1988), pp. 61.

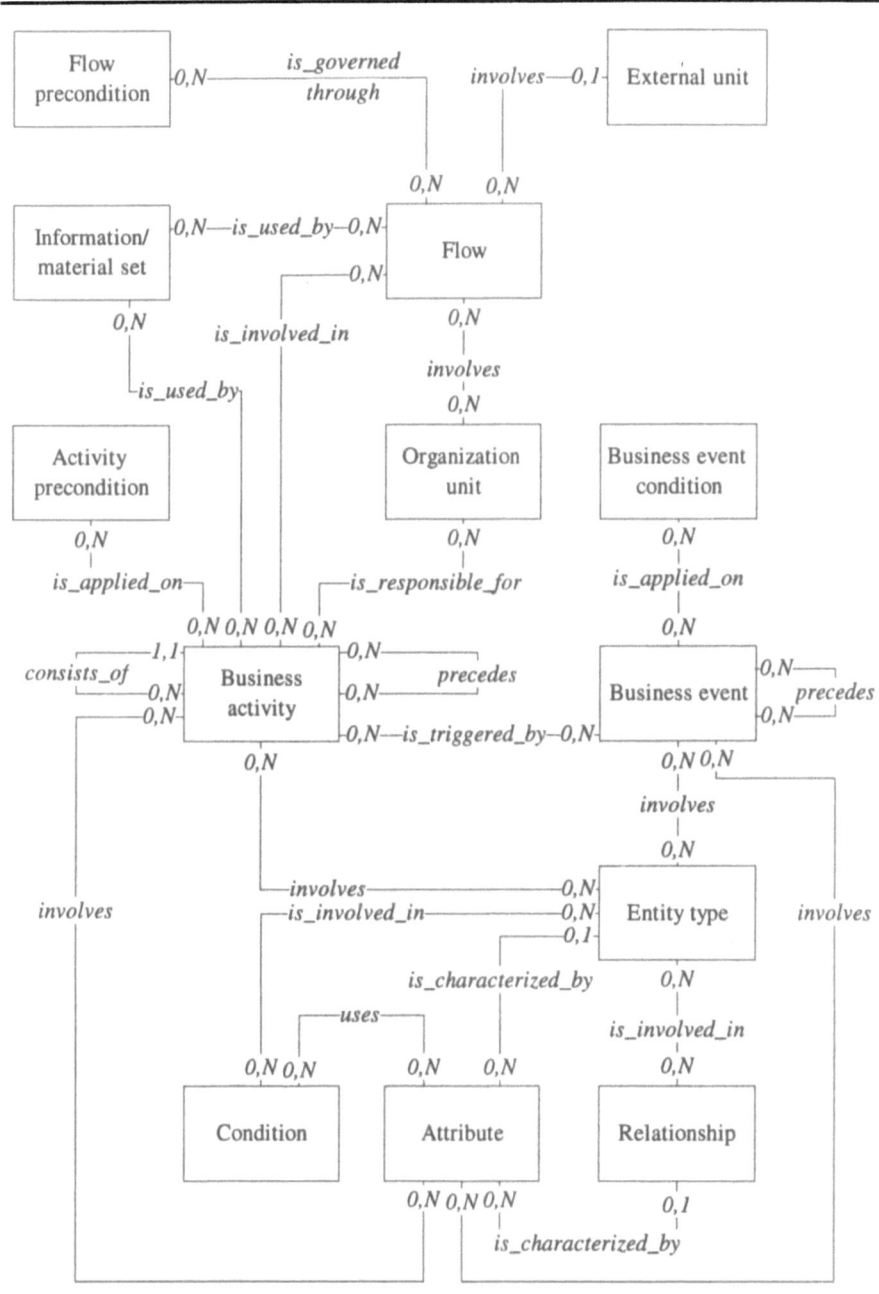

Figure 4-9: Meta model of IFIP WG 8.1[23]

23 Olle et al. (1988), p. 61 (data perspective), p. 78 (process perspective), p. 90 (behavior perspective), and p. 94 (cross-reference perspective).

To evaluate the meta model, the representation of business rules, processes, data models, and organizational facts is regarded:

- *Business rules*: Regarding ECA^2 rules, the meta model of the IFIP WG 8.1 encompasses meta entity types for the specification of the basic components event, condition, and action. While events and actions can be seen as semantically equivalent to the constructs of the business rule approach, the condition component is treated differently. Events in the meta model are governed by post- or preconditions. Postconditions can be regarded as equivalent to the condition component of business rules, whereas the preconditions are an additional construct of the IFIP WG 8.1 meta model. Another condition construct relevant with respect to business rules is the *precondition* related to business activities which could also be assigned to the condition component of ECA^2 rules. This definition of two separated condition constructs for business events and business activities may lead to redundancies as it e.g. happened within the examples given by Olle et al.:

  > *Tuple in the meta entity type* Activity precondition*:*
  > *(#2, 'Stock below reordering level')*[24].
  >
  > *Tuple in the meta entity type* Business event condition:
  > *(#1, 'Stock below reordering level')*[25]

  The existence of such redundancies may lead to maintenance problems and could be avoided by e.g. using a single meta entity type as done in the business rule meta model.

- *Processes*: The modeling of processes is primarily supported by the meta entity type *Business activity* using the two recursive relationship types *consists of* and *precedes*. The first business activity of a process is triggered by at least one business event; however, within the process, the activities are explicitly linked resulting in rather fixed sequences. Thus, the potential of events has not fully been exploited.

- *Data model*: The IFIP WG 8.1 meta model provides constructs for the specification of an ERM. These are linked by several *involve* relationships to *Business event, Business activity,* and *Condition* thus providing a transparency of data use within the meta data stored in these meta entity types.

- *Organizational facts*: In the meta model regarded, organization units, process responsibilities, and material and information flows are considered. The responsibility for specific processes is assigned to an organization unit. However, closely related organizational facts like manual/automated processors and organizational structures are omitted.

---

[24] Olle et al. (1988), p. 86.
[25] Olle et al. (1988), p. 93.

4.1.4.1.2 ESPRIT: CIM Open Systems Architecture

Computer integrated manufacturing (CIM) attempts to integrate the information exchange within a manufacturing company with computer aided tools, e.g., CAD or CAM. To achieve this goal, *generalized* concepts had to be developed which identified the principal components, processes, constraints, and information sources needed to describe a CIM environment. A research project of the European Community was launched 1986 and carried out by the ESPRIT Consortium AMICE[26]. The goals of the project were to develop an Open System Architecture for CIM (CIM-OSA) which should provide

- a general definition of the scope and nature of CIM,
- guidelines for implementation,
- a description of constituent systems and subsystems, and
- a modular framework complying with international standards.

As a conceptual framework, CIM-OSA is based on a meta model which encompasses a function view (cf. Figure 4-10) and an information view (cf. Figure 4-11). The roles of the relationship types included in the meta models have to be read in the direction indicated by the arrows.

---

[26] ESPRIT Consortium AMICE (1989).

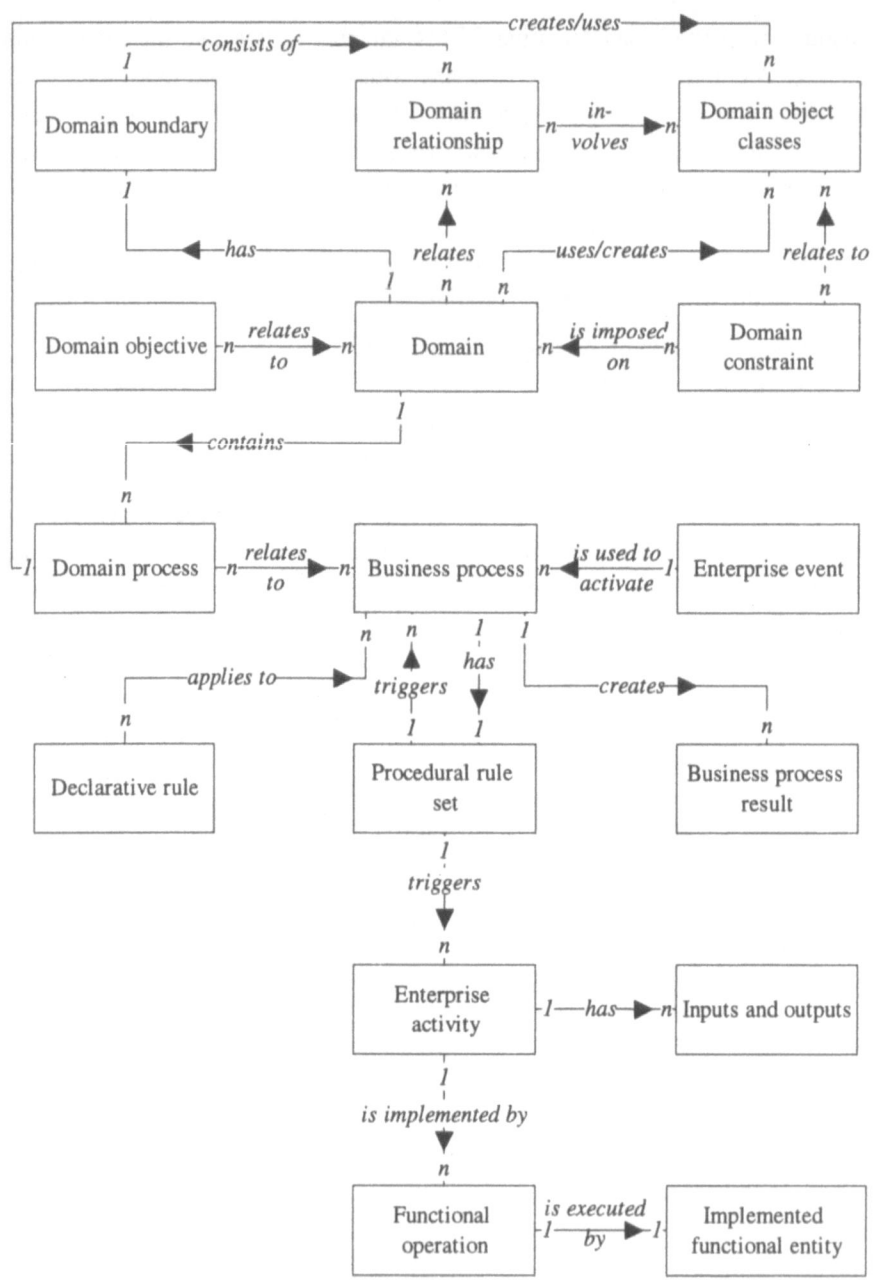

Figure 4-10: Function view concepts[27]

27   Jorysz/Vernadat (1990a), p. 151.

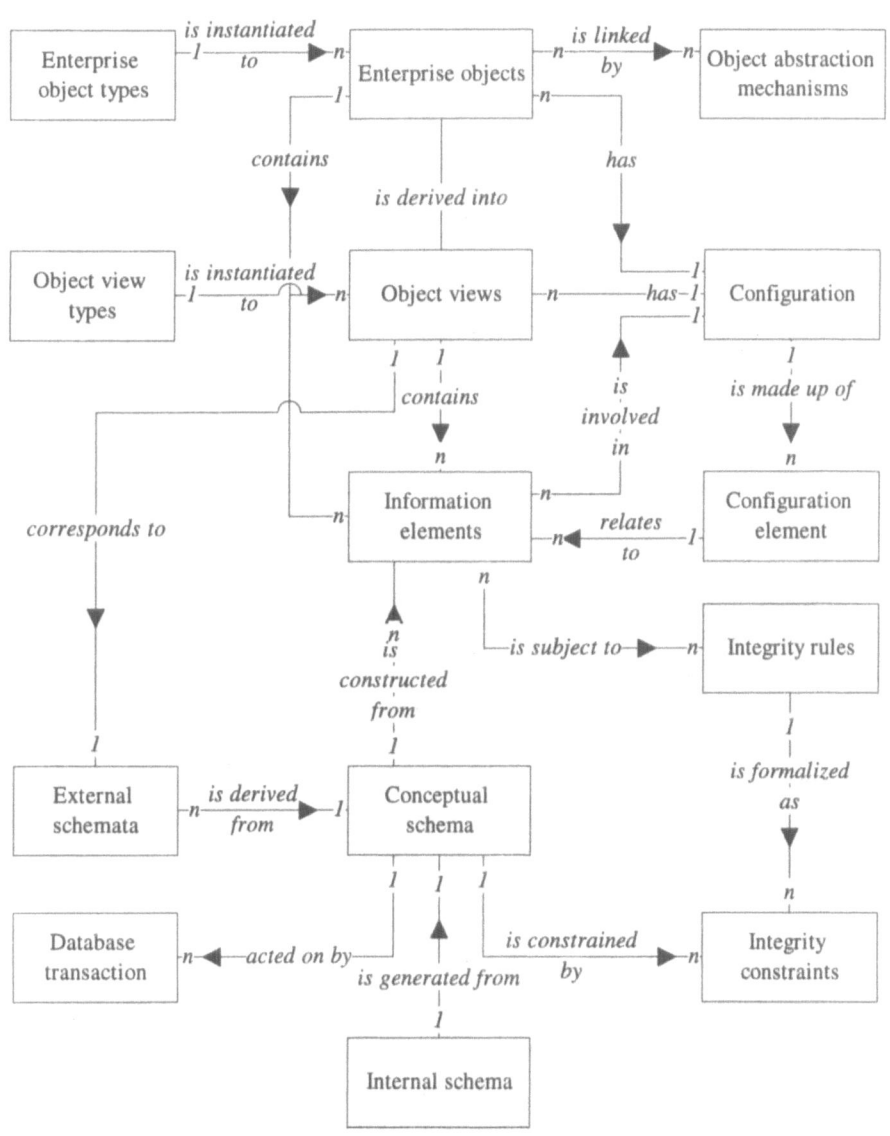

Figure 4-11: Information view concepts[28]

These two views of the meta model are analyzed in the following with respect to the semantics of ECA^2 rules, processes, data model, and organizational facts.

- *Business rules*: The function view of the meta model encompasses a meta entity type *Enterprise event* which is related to *Business process*. Therefore,

---

[28] Cf. Jorysz/Vernadat (1990b), p. 159.

each business process is invoked by exactly one enterprise event which may trigger several processes.

Conditions are specified using one of the two meta entity types *Domain constraint* or *Integrity constraints*. However, conditions restricting function executions, i.e., automation rules are not explicitly considered, but included in enterprise events.

Based on this meta model, actions can be specified using the meta entity types *Domain process, Business process*, or *Enterprise activity* which are elements of the functional view. A business process may encompass several domain processes and enterprise activities. Thus, for the specification of an action, one has to classify it in one of those types. However, there may exist situations in which this classification is ambiguous.

The meta model of CIM-OSA encompasses a meta entity type *Procedural rule set* which could be regarded as equivalent to the construct of business rules. The procedural rule set „defines the desired sequence of the enterprise activities and/or business processes in the form of a flow of control"[29]. However, this quote indicates that the scope of these rules is narrower than the one of business rules because they specify automation rules which are only a subtype of business rules[30].

- *Processes*: In the meta model of CIM-OSA business processes are defined by a set of seven related constructs, some of them appearing as meta entity types:
  - start conditions (including enterprise events),
  - a set of business processes and enterprise activities it uses,
  - a set of procedural rules specifying the behavior (see above),
  - a set of declarative rules (i.e. constraints),
  - a set of finish conditions, and
  - the intended result of the business process[31].

  Compared to the process specification by using business rules, a much larger heterogeneity of the specification is obvious. In the business rule approach, the start of a process as well as its behavior is described with ECA^2 rules. Thus, fewer modeling constructs are needed and as already mentioned in the discussion of the IFIP WG 8.1 meta model, the behavior specification based on raised and triggering events is more flexible than a 'hard-coded' sequentialization of activities.

- *Data model*: Because CIM-OSA is intended to be a general approach, its meta models do not incorporate meta entity types of a specific conceptual

---

[29] Jorysz/Vernadat (1990a), p. 153.

[30] Cf. Chapter 3.

[31] Cf. Jorysz/Vernadat (1990a), pp. 152.

data model. Because of the separation of the meta model into views, the relationship between data model constructs and functional specifications is not transparent. This could be achieved by an integration view like in the IFIP WG 8.1 meta model.

- *Organizational facts*: Organizational facts are included in the meta entity type *Domain* which is defined by objectives, processes, object classes, and boundaries to other domains. However, as Jorysz and Vernadat state „it is very important to mention that a domain is not specified by a subset of the organization chart of an enterprise, but depends on a given set of business objectives and constraints to be satisfied"[32]. In the CIM-OSA meta model, organizational facts are only related to processes, whereas relationships to elements of the information view are omitted.

4.1.4.1.3   ARIS

The ARIS approach[33] focuses on the modeling of business processes and is often used in relation with the introduction of the enterprise management system SAP R/3[34]. The meta model of ARIS consists of several closely interrelated views:

- The *data view* is defined by events and states.
- The *functional view* encompasses the description of functions and their interdependencies.
- The *organizational view* deals with the structure of organizational units and the assignment of employees.
- The *resource view* is defined by the components of the information technology.

Although the separation of these views reduces the complexity, it makes their interdependencies intransparent. Therefore, the integration of the first three views is provided by a *control view* as depicted in Figure 4-12. Within this framework three abstraction layers are distinguished[35]. In the context of systems analysis, the conceptual model (semantic model) is of interest and only this part of the ARIS meta model is considered in the following[36].

---

[32]   Jorysz/Vernadat (1990a), pp. 151.
[33]   Cf. to the following Scheer (1995), pp. 11.
[34]   Cf. Keller (1994).
[35]   Cf. Scheer (1995), pp. 14.
[36]   For detailed information on the other layers cf. Scheer (1995).

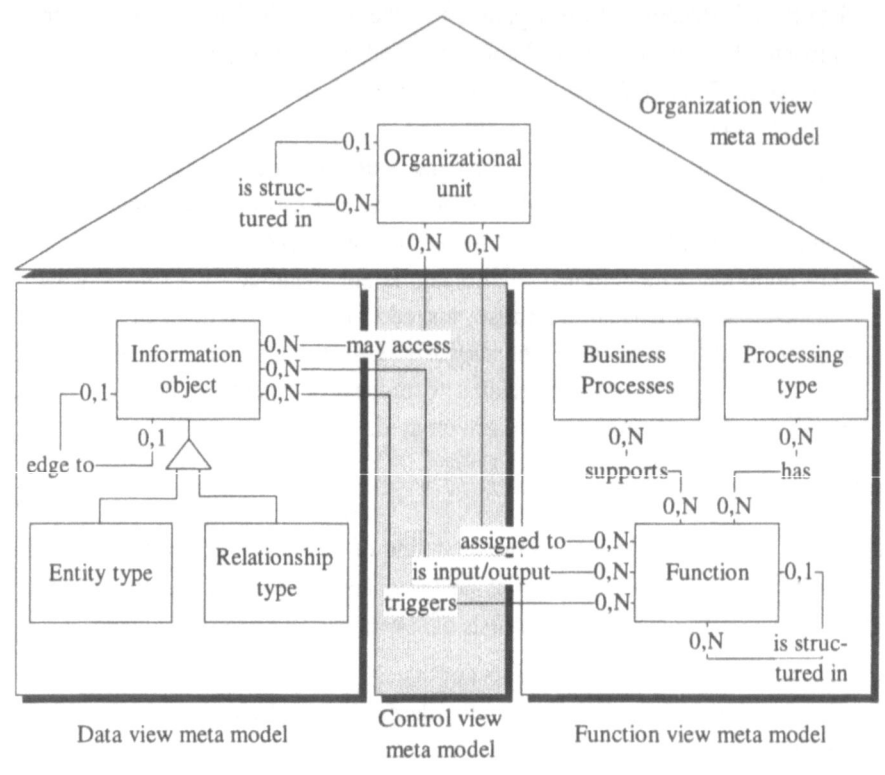

Figure 4-12: Overview on the meta models of ARIS[37]

Though ARIS encompasses a detailed model for organizational facts[38], this model is not further considered because the focus of the discussion is the representation of business rules. In ARIS the meta model for the data structure is very detailed and precise; however, it is not depicted here because it also encompasses the meta entity types for the specification of an ERM[39]. Thus, the presentation focuses on the *function* and the *control meta model*.

The meta model for the *function view* encompasses the meta entity types for the specification of processes and includes constructs for the definition of a decision model and of enterprise goals (cf. Figure 4-13). Relationship types without role names result from a modification of complex relationships.

---

[37] Cf. Scheer (1995), p. 701.
[38] The meta model for the organizational structures is depicted in Scheer (1991), p. 92.
[39] The meta model for the data model is depicted in Scheer (1991), p. 104.

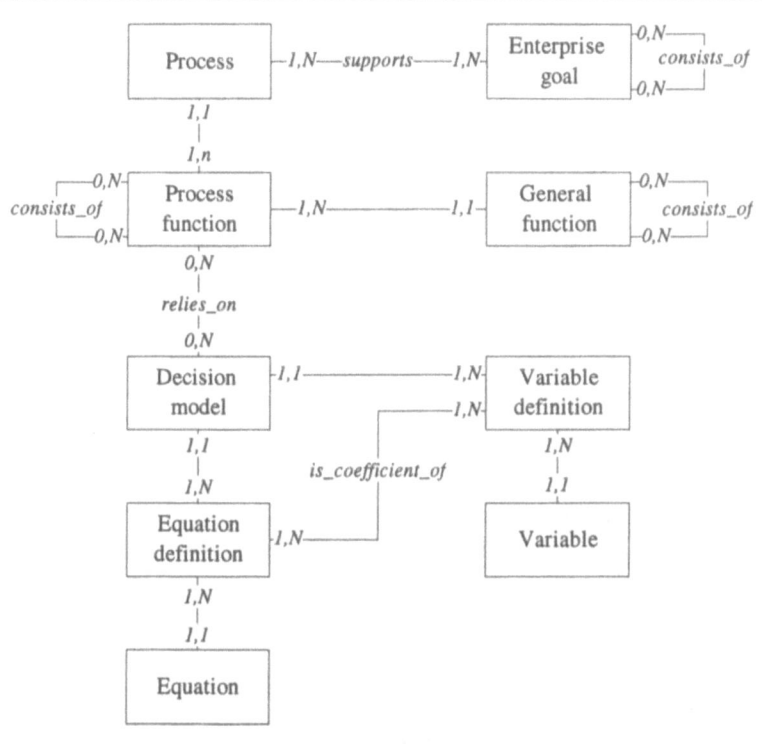

Figure 4-13: Meta model for the function view of ARIS[40]

The meta model for the *control view* encompasses the meta entity type *Event* as a subtype of *Information object*. Events may trigger functions and this triggering refers to attributes of entity types[41] (cf. Figure 4-14). Furthermore, events in ARIS are the result of functions. The analysis of both relationships facilitates the description of event-oriented process chains as a type of high level Petri nets[42].

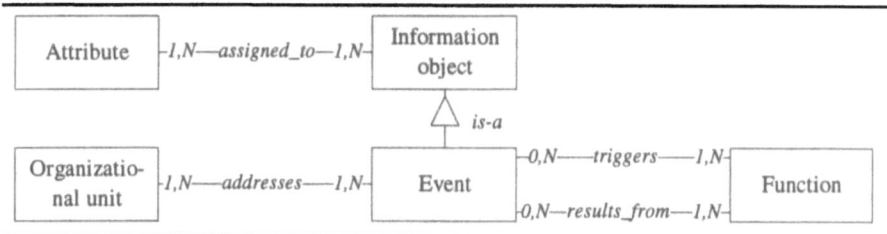

Figure 4-14: Events in ARIS[43]

---

[40] Scheer (1991), p. 79.
[41] Cf. Scheer (1991), p. 119.
[42] Cf. Scheer (1991), pp. 130; Hoffmann et al. (1993).
[43] Model derived from Scheer (1991), pp. 114.

Similar to the meta models presented above, the ARIS meta models are evaluated
with respect to business rules, processes, data model, and organizational facts:

- *Business rules*: Though the meta entity type *Event* is not emphasized in the
  meta model and the semantics of events in ARIS remain rather ambiguous,
  events are very important elements of the approach[44].
  ARIS encompasses a decision model in the function view, but does not provide a meta entity type for the explicit specification of the condition component of business rules. Such a meta entity type would have to be allocated
  between the meta entity types *Event* and *Function*.
  The action component of business rules can be related to the meta entity
  type *General function* of ARIS.
- *Processes*: The specification of processes is done using event-oriented process chains which result from the two relationships between events and
  functions. In this property ARIS is similar to the business rules approach,
  i.e., the event-oriented process chain is a graphical model which could also
  be used for meta data stored in the business rule meta model.
- *Data model*: The data model used in ARIS is an extension of the ERM.
  Relationships between submodels are considered in the meta model, providing an integrated model of the universe of discourse.
- *Organizational facts*: ARIS considers a large number of organizational
  facts, e.g., organizational structures, positions, employees, and decision
  models which are related to functions.

4.1.4.1.4   RIM

The meta model discussed in the following has been developed in the RIM department of the Institute of Information Systems, University of St. Gall[45]. The
goal of the model is to provide a framework for the evaluation of widely used
software development methods[46].

---

[44]   Cf. Hoffmann et al. (1993).

[45]   RIM is an acronym for '*Rechnergestütztes Informationsmanagement*' (computer-based
       information management).

[46]   Cf. Färberböck et al. (1991), pp. 41.

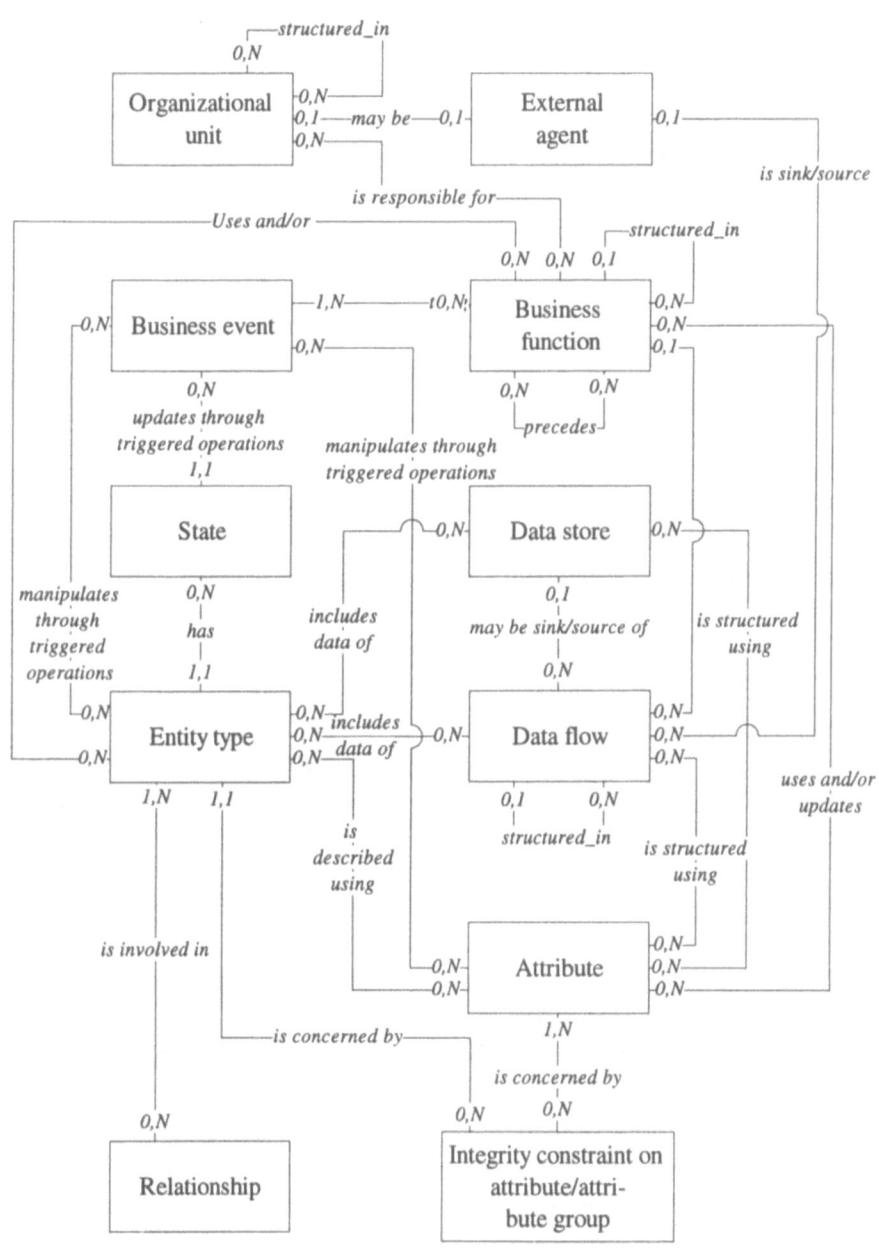

Figure 4-15: Analysis meta model of Österle/Gutzwiller[47]

---

[47] The meta model is an aggregation of the models depicted in Österle/Gutzwiller (1992a), pp. 47.

Figure 4-15 depicts the reference meta model of RIM[48] which is shortly discussed below with respect to the representation of business rules, processes, components of the data model, and organizational facts:

- *Business rules*: For the specification of events, the meta entity type *Business event* exists. This meta entity type has relationships to four other meta entity types:
  - *Business function*: A business function is triggered by an event.
  - *State*: Operations triggered by an event may change the state (of an entity).
  - *Entity type*: Operations triggered by an event may directly change an entity.
  - *Attribute*: Operations triggered by an event may manipulate an attribute.

  Only conditions describing integrity constraints can be explicitly specified (using the meta entity type *Integrity constraint on attribute/attribute group*). Other conditions which may be expressed by the condition part of a business rule (e.g. preconditions for the execution of actions) are part of the relationship types between the meta entity types *Business event* and *Entity type, Attribute,* or *Business function,* respectively.

  A subset of possible actions of a business rules is explicitly expressed using the meta entity type *Business function.* However, similar to conditions, 'lower' level operations which probably only manipulate data objects are only implied by the relationship types between *Business event* and data model components.

- *Processes*: The meta entity type *Business function* and the two recursive relationships *consists of* and *follows* may be used to specify process hierarchies and processes as sequences of functions. However, these function sequences are again 'hard-coded' and not dynamically established by events as in the business rule approach.

- *Data model*: The data model supported by the meta model is the standard ERM. There exist relationship types between the components of the data model and the two meta entity types *Business event* and *Business function* which indicate manipulations triggered by an event. However, the component of a data model specifically involved in raising an event can not be represented.

- *Organizational facts*: The meta model encompasses the meta entity types *Business function, External agent,* and *Business unit* which are relevant for a comprehensive model of the universe of discourse.

---

[48] Cf. also Heym/Österle (1992).

4.1.4.2 Comparison of the Selected Meta Models

The *BROCOM* meta model encompasses business rules as central modeling constructs which are linked to closely related facts. In the previous subsections, several selected meta models of other approaches were presented and shortly compared with the *BROCOM* meta model. This is summarized in Figure 4-16.

	BROCOM	IFIP WG 8.1	CIM-OSA	ARIS	RIM
			Meta model of		
Event	Elementary and composite events	Business event	Enterprise event	Subtype of meta entity type *information object*; relationship between function and information objects	Business event
Condition	Condition	Pre- and post-condition on business events, precondition on business activities	Domain constraint for processes; integrity constraint	Decision model, integrity constraints	Integrity constraint; implicit relationship between business events and data model components.
Action	Action	Business activities	Business process; domain process; enterprise activity	Business process, function	Business function; implicit relationship between business events and data model components.
Data model	ERM, NIAM	ERM	Generic; no restriction	ERM	ERM
Organizational facts	Process structure; process; organization unit and structure; processors; origin	Organization unit; responsibility for executing business activities; material and information flow.	Domain (additional elements are part of the organization view which is not considered in this work)	Organization unit; organizational structure; positions, employees	Business unit; external agents (results from including data flow diagrams)

Figure 4-16: Comparison of selected meta models

4.2   Steps of the *BROCOM* Approach

In systems analysis, a main goal is to collect all information about the universe of discourse relevant to the development of an IS[49]. These facts are primarily about processes and about the structure and manipulation of data. Their collection may involve methodologies from several areas, e.g., marketing, sociology, psychology, and also information engineering; however, collecting knowledge always is confined by limits to what *can* be known (because of theoretical and practical reasons) and limits to what *should* be known (because of ethical and social reasons)[50]. In the following, the steps of the *BROCOM* approach are presented.

The specification of the universe of discourse relevant for a large IS may lead to an immense amount of business rules. To reduce this complexity, two approaches are applied:

- Top down specification of processes in hierarchies of super- and subprocesses and separate behavior descriptions for each subprocess.
- Distinction between business rules that are explicitly assigned to at least one (sub-) process and rules that are valid for all (sub-) processes (cf. Figure 4-17). The first set of rules is registered during the specification of processes and mainly consists of automation rules and only of very few integrity rules. A large part of the rules of the second set are integrity rules which are registered during the completion of the conceptual data model.

Figure 4-17: Two different sets of rules in BROCOM

---

[49] Cf. Pohl (1993).
[50] Jones/Walsham (1992), pp. 196.

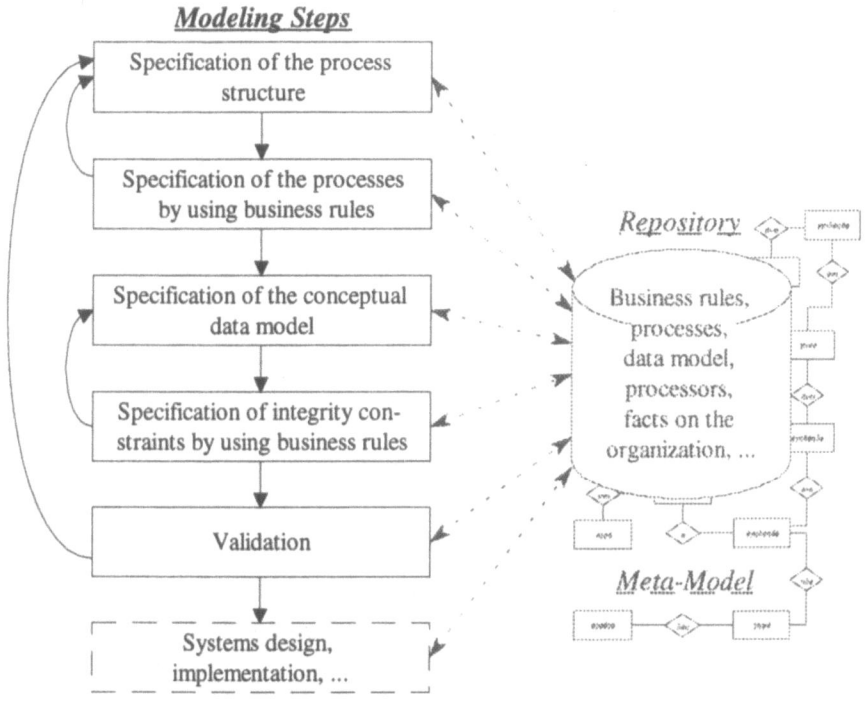

Figure 4-18: Framework of **BROCOM:** *Modeling steps*

From this distinction, the first four steps of **BROCOM** result. Steps one and two concern *process specific business rules* and steps three and four *process independent business rules* (cf. Figure 4-18). In the following, all five steps are shortly explained and specified by using business rules:

1. *Process structure:* The process structure provides an overview of the process interdependencies within the universe of discourse and allows for the division of processes in modular and thus reusable and maintainable subprocesses.

    Business Rule [SA-1] 'PROCESS-STRUCTURE'
    *ON   (**start conceptual modeling**)*
    THEN specify process-structure;
            *raise event '**PROCESS-STRUCTURE-SPECIFIED**'*

2. *Process dynamics:* In this step processes and sub-processes, i.e., the nodes of the process structure are specified by using business rules. Within this task, business rules can be further refined, which may extend the process structure. Therefore, the processes in the structure are always a superset of those specified using business rules.

Business Rule [SA-2] 'PROCESS-DYNAMICS'
  ON   **(process structure specified)**
  IF   *(process structure sufficient)*
  THEN   *specify processes using business rules;*
          raise event **'PROCESSES-SPECIFIED'**
  ELSE   *issue error message to analyst;*
          raise event **'START-OF-SYSTEMS-ANALYSIS'**

3. *Conceptual data model:* After having described the processes, all data model components used within processes can be derived from the business rules and are structured in the conceptual data model.

Business Rule [SA-3] 'DATA-MODEL'
  ON   **(processes specified)**
  THEN   *specify conceptual data model;*
          raise event **'CONCEPTUAL-DATA-MODEL-SPECIFIED'**

4. *Integrity constraints:* In the second step mostly automation rules have been described and assigned to processes. Now the conceptual data model is analyzed with regard to static and dynamic integrity constraints which are specified as business rules[51].

Business Rule [SA-4] 'INTEGRITY-CONSTRAINTS'
  ON   **(conceptual data model specified)**
  THEN   *specify integrity constraints by using business rules;*
          raise event **'CONCEPTUAL-MODEL-COMPLETE'**

5. *Validation:* The completed conceptual model has to be validated by analysts and clients (e.g. end users).

Business Rule [SA-5] 'VALIDATION'
  ON   **(conceptual model complete)**
  IF   *(conceptual model invalid)*
  THEN   *correct errors;*
          raise event **'CONCEPTUAL-MODEL-COMPLETE'**
  ELSE   *confirm validity;*
          raise event **'CONCEPTUAL-MODEL-VALIDATED'**

Finally, the verified and validated conceptual model may be used as a basis for systems design and for the implementation by using e.g. the alternatives discussed in Chapter 2. As an overview of the interdependencies between the meta model and the modeling steps, Figure 4-19 shows in which step the submodels are manipulated.

Task	Submodels of the meta model					
	Business rule	Process	Data model	Processor	Org. unit	Origin
Process structure		W				
Process specification	W	R/W		W	W	W
Conceptual data model	R		W			
Integrity constraints	R/W		R	R/W	R/W	R/W
Validation	R/W	R/W	R/W	R/W	R/W	R/W

Figure 4-19: Interdependencies between tasks and meta entity types (W: Write, R: Read)

In the following sections, these steps of *BROCOM* are presented in detail. As already done above, each step of the approach is specified using *business rules*.

4.2.1   Step 1: Process Structure

4.2.1.1   Deliverables

The process structure provides a first overview on processes and subprocesses relevant within the universe of discourse. In order to have modular and reusable processes, each subprocess may have several preceding processes. Thus, the resulting structure is not a tree but a hierarchical network of process nodes. The number of levels in the structure depends on the precision attempted; however, it can always be extended during the next modeling steps.

4.2.1.2   Procedure

Applying the principles of systems engineering, a top-down approach is chosen in which the process structure is developed by starting with a top-level process and refining it at lower levels. By using business rules, the step of defining the *process structure* is specified as follows:

Business Rule [SA-1-1] 'PROCESS-STRUCTURE'
>   *ON*   *(start conceptual modeling)*
>   *THEN*   *register top-level-process;*
>   *raise event 'SUBPROCESSES-TO-BE-SPECIFIED'*

Business Rule [SA-1-2] 'REGISTER-SUBPROCESSES'
>   *ON*   *(subprocesses to be specified)*
>   *THEN*   *register subprocesses;*
>   *raise event 'SUB-PROCESSES-SPECIFIED'*

---

[51]   This step is also described by Appleton (1986), pp. 88; however, Appleton restricts the term of business rules to integrity rules.

Business Rule [SA-1-3] 'CONCLUDE-PROCESS-STRUCTURE'
    ON   **(subprocesses specified)**
    IF   *(further subprocesses necessary)*
    THEN   select next process node;
              raise event **'SUBPROCESSES-TO-BE-SPECIFIED'**
    ELSE   confirm process structure;
              raise event **'PROCESS-STRUCTURE-SPECIFIED'**

4.2.1.3 Application to DOM-2

The application of the modeling task *process structure* to *DOM-2* results in a
main process *damage administration* which is divided in the four subprocesses
damage reception, damage registration, damage treatment, and *damage archiving* (cf. Figure 4-20).

- The subprocess *damage reception* includes the handling of a damage claim
  prior to its definitive registration in the DOM-2 system.
- In the subprocess *damage registration* all data relevant to the damage administration is registered. The goal of this subprocess is to check whether a
  damage is covered by a policy of the policy holder.
- The subprocess *damage treatment* contains the payments and is only invoked if the damage claim has been accepted.
- The subprocess *damage archiving* encompasses the archiving of all documents of the damage.

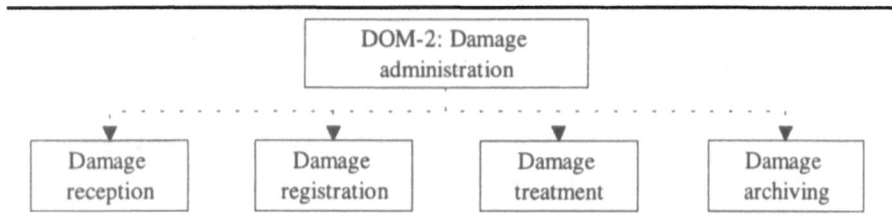

Figure 4-20: Process structure of DOM-2

4.2.2 Step 2: Process Specification

4.2.2.1 Deliverables

The process structure developed in the preceding step is only a static view on
processes in the universe of discourse and does not imply any temporal orders. In
the step *process specification,* the content of all relevant nodes, i.e., the behavior

the description of all processes, each component of the registered ECA[2] rules is assigned to origins, processors, and (indirectly) organizational units.

The model which results from this modeling step reveals

- process dynamics,
- interfaces between manual and automated process execution,
- interfaces between organizational units resulting from the assignment of processors to organizational units, and
- origins of business rules and thus the origin owners who are responsible for the specified business rules.

This model can be used e.g. for designing and evaluating process alternatives.

4.2.2.2 Procedure

Similar to the definition of a process structure, a specification of processes begins with the top-level process and is carried out using a top-down approach. A refinement at a lower level is always done based on a single business rule which is linked to a subprocess. This is also done in the business rule based description of the modeling tasks: business rule [SA-1] 'PROCESS-STRUCTURE' is refined by the subprocess *process structure* which itself is described by business rules [SA-1-1] through [SA-1-3]. In the following, the business rule [SA-2] 'PROCESS-DYNAMICS' is refined by the subprocess *process specification* containing business rules [SA-2-1] through [SA-2-7].

Business Rule [SA-2-1] 'CHECK-PROCESS-STRUCTURE'

ON	*(process structure specified)*
IF	*(process structure sufficient)*
THEN	*select a top-level process to specify;*
	raise event **'PROCESS-TO-SPECIFY'**
ELSE	*issue error message to analyst;*
	raise event 'START-OF-SYSTEMS-ANALYSIS'

Business Rule [SA-2-2] 'START-OF-PROCESS'

ON	*(process to specify)*
THEN	*specify first business rule with correct starting event;*
	raise event **'PROCESS-START-SPECIFIED'**

Business Rule [SA-2-3] 'SPECIFY-PROCESS'

ON	*(process start specified)*
THEN	*specify rest of process by using business rules;*
	raise event **'PROCESS-PROV-SPECIFIED'**

Business Rule [SA-2-4] 'CONFIRM-SUBPROCESS'

    ON   **(process provisionally specified)**

    IF   *(triggering event = triggering event of refined business rule) AND*
                  (raised events = raised events of refined business rule)

    THEN   *confirm specification of the process;*
                  raise event **'PROCESS-COMPLETELY-SPECIFIED'**

    ELSE   *correct process specification;*
                  raise event **'PROCESS-SPECIFIED'**

Business Rule [SA-2-5] 'CONCLUDE-PROCESS-SPECIFICATION'

    ON   **(process specified)**

    IF   *(all relevant processes specified)*

    THEN   *confirm process specification;*
                  raise event **'PROCESSES-SPECIFIED'**

    THEN   *continue process specification;*
                  raise event **'CONTINUE-PROCESS-SPECIFICATION'**

Business Rule [SA-2-6] 'CHOOSE-NEXT-PROCESS'

    ON   **(process specification to be continued)**

    IF   *(refinement of current process necessary)*

    THEN   *start refinement in current process;*
                  raise event **'BUSINESS-RULE-TO-REFINE'**

    ELSE   *choose next process;*
                  raise event **'BUSINESS-RULE-TO-REFINE'**

Business Rule [SA-2-7] 'REFINE-BUSINESS-RULE'

    ON   **(business rule to be refined)**

    THEN   *choose an unrefined business rule of the current process;*
                  link chosen business rule to a subprocess of the current process;
                  raise event **'PROCESS-TO-SPECIFY'**

The refinement of business rules described in rule [SA-2-7] is further shown in Figure 4-21: A business rule of the current process is selected for refinement (Figure 4-21a) and afterwards assigned to a subprocess of the current process. In Figure 4-21b, *process 2* is chosen. Finally, this linked subprocess is specified by using business rules (cf. Figure 4-21c).

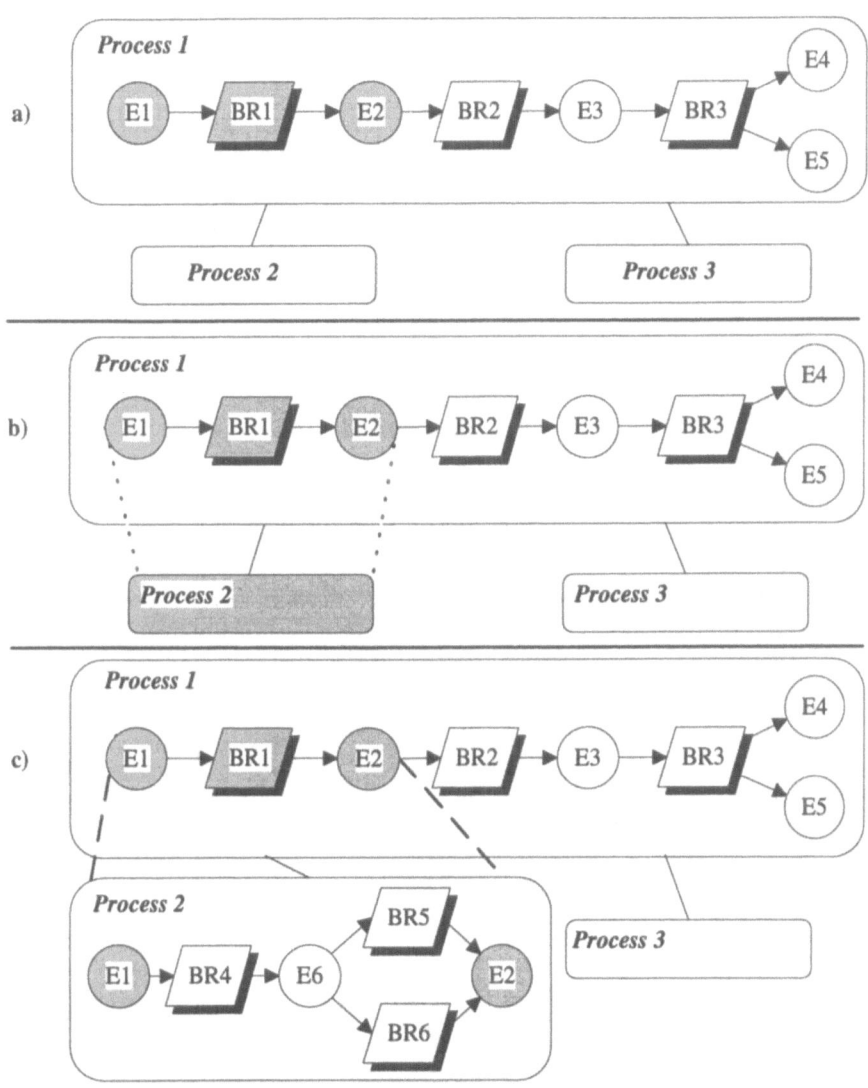

Figure 4-21: Refinement of [BR-1] by process 2

After the complete specification of all process nodes relevant for the IS, the modeling step *process specification* can be concluded and the next step, the definition of the conceptual data model, may be initiated.

As emphasized before, the registration of all business rules has to be done by using a repository system. However, in working with end users, a previous manual registration of business rules may be appropriate. For this, a form like the one depicted in Figure 4-22 may be applied in order to formalize business rules in the ECA^2 structure:

- The *form header* includes information on the analyst/end user filling in the form, the date of the registration, and a form number.
- The *rule header* encompasses a name, a short, unstructured description of the rule content, and the origin from which the rule has been retrieved.
- The *rule body* consists of the four ECA^2 elements of a business rule.
- The *rule footer* lists the processors who/which currently execute the rule components.

| Registrator *Herbst Holger* | Date *15-09-96* | No. 254 |

Rule-Name *Product-Reorder*

Description *The rule defines that a product has to be automatically reordered if the stock is lower than a threshold*

Origin *Stockkeeping guideline, p. 49*

Event *Stock is reduced*

Condition
Stock < Threshold

Then action
Order product at our supplier

Else action

Processors
E: *Stock keeper*
C: *Stock keeper*
A: *Purchaser*
A:

Figure 4-22: Registration form for business rules

In this form, end users may use their syntax for describing a business rule; however, the rule formalization is at least done with respect to the ECA^2 structure. Because the automatic identification of the referred data model components has supported in the next modeling step, they may already get marked in the form (e.g. by underlining them). After the manual registration, the forms are collected and the facts registered in the repository.

4.2.2.3   Application to DOM-2

The process structure of DOM-2 encompasses *damage administration* (DA) as top-level process. This process is initiated by the disjunction of the events *phone call of person* and *letter of person*. The specified process finally consists of four business rules:

Business Rule [DA-1] 'DAMAGE-RECEPTION'
    ON   **(phone call of person) OR (letter of person)**
    IF   (definite damage registration needed)
    THEN   start damage registration;
             raise event **'DEF-DAMAGE-REGISTRATION'**
    ELSE   reject damage;
             raise event **'DAMAGE-REJECTED'**

Business Rule [DA-2] 'DAMAGE-REGISTRATION'
    ON   **(damage to be registered)**
    IF   (damage probably covered by policy)
    THEN   register damage;
             raise event **'DAMAGE-DEFINITELY-REGISTERED'**
    ELSE   reject damage;
             raise event **'DAMAGE-REJECTED'**

Business Rule [DA-3] 'DAMAGE-TREATMENT'
    ON   **(damage definitely registered)**
    IF   (damage covered by policy)
    THEN   process damage claim;
             raise event **'DAMAGE-CLOSED'**
    ELSE   reject damage claim;
             raise event **'DAMAGE-REJECTED'**

Business Rule [DA-4] 'ARCHIVE-DAMAGE'
    ON   **(damage closed)**
    THEN   archive damage;
             raise event **'DAMAGE-ARCHIVED'**

In the following, the business rule [DA-1] is refined by the subprocess *damage reception* that consists of the six business rules which have already been used as example rules in the preceding chapter. Therefore, only those rules are repeated which raise events that are elements of the refined business rule of the upper-level process:

Business rule [DA-1-1] 'PERSON-CONTACTS-US'
    ON   **(phone call of person) OR (letter of person)**
    IF   *(person is a policy holder) AND (policy holder reports damage)*
    THEN *begin damage registration;*
             raise event 'DAMAGE-REPORTED'

Business rule [DA-1-2]
    ON   ...*

Business rule [DA-1-3] 'ACCEPT-PROV-DAMAGE'
    ON   *(provisional damage registered)*
    IF   *(damage covered by policy)*
    THEN *accept damage provisionally;*
             raise event 'DAMAGE-PROV-ACCEPTED';
             send damage form to policy-holder;
             raise event 'DAMAGE-FORM-SENT'
    ELSE *reject damage;*
             *raise event **'DAMAGE-REJECTED'***

Business rule [DA-1-4]
    ON   ...*

Business rule [DA-1-5]
    ON   ...*

Business rule [DA-1-6] 'DEF-REGISTER-DAMAGE'
    ON   *(DAMAGE-FORM-RECEIVED) WITHIN (90 DAY AFTER*
             (DAMAGE-FORM-SENT))
    IF   *(information is complete)*
    THEN *raise event **'START-DEF-DAMAGE-REGISTRATION'***
    ELSE *return form to policy-holder;*
             raise event 'DAMAGE-FORM-SENT'

Because the specification of processes may result in the need for further refinement levels, the process structure may be extended during the modeling task *process specification.* Figure 4-23 encompasses four subprocesses of *damage registration* which have not been specified in Figure 4-20. Furthermore, the diagram now reveals additional information on the current state of process specification:

- Process nodes *without shadow* are not specified, i.e., they do not contain any business rules.
- Process nodes depicted as *rectangles with shadow* are currently being specified.
- Process nodes with *rounded corners and shadow* are completely specified.

- *Dotted lines* are connections which have not yet been assigned to a refined business rule of the higher-level process.
- *'Normal' lines* are connections which are assigned to a specific business rule of the higher-level process.

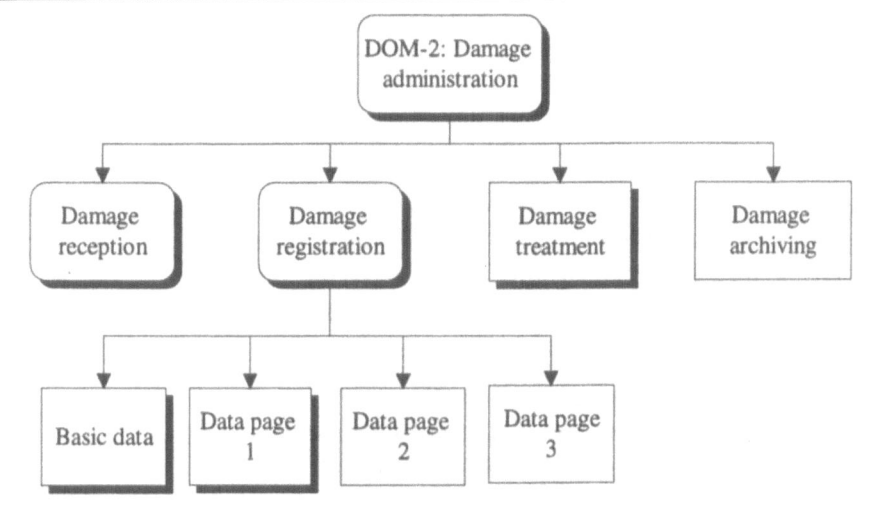

Figure 4-23: Process structure after the step process specification

4.2.3   Step 3: Data Model

4.2.3.1   Deliverables

The modeling task *data model* results in a conceptual data model which may be specified by the means of e.g. ERM or NIAM[53]. The conceptual data model consists of at least those components which are referred by the business rules which constitute the processes.

4.2.3.2   Procedure

The analysis of business rules with respect to components of the data model results in an unstructured set of constructs which have to be manipulated in order to get a correct conceptual data model. The derivation of data model components from the content of the business rules has to be supported by keywords within the rule formalization. For the derivation of components of the ERM the following keywords may be used:

---

[53] NIAM is an acronym for Nijssen's Information Analysis Methodology.

- Entity type: ⇨ENT
- Relationship type: ⇨REL
- Attribute: ⇨ATT

Using again the notation of business rules, this task of a business rule-oriented conceptual modeling is described as follows:

Business Rule [SA-3-1] 'DERIVATION'

    ON   (processes specified)

    THEN derive all components of the data model;
              raise event 'DATA-MODEL-COMPONENTS-DERIVED'

Business Rule [SA-3-2] 'RELATIONSHIP-TYPES'

    ON   (data model components derived)

    THEN link entity types by relationship types;
              raise event 'RELATIONSHIP-TYPES-ESTABLISHED'

Business Rule [SA-3-3] 'ATTRIBUTES'

    ON   (relationship types established)

    THEN assign attributes to entity types and relationship types;
              raise event 'ATTRIBUTES-ASSIGNED'

Business Rule [SA-3-4] 'CONCLUDE-DATA-MODEL'

    ON   (attributes assigned)

    IF   (conceptual data model consistent)

    THEN confirm conceptual data model;
              raise event 'DATA-MODEL-SPECIFIED'

    ELSE correct errors;
              raise event 'ATTRIBUTES-ASSIGNED'

4.2.3.3 Application to DOM-2

The use of the keywords ⇨ENT, ⇨REL, and ⇨ATT in the formalization of the business rules which specify the processes of DOM-2 allows for the automated derivation of the following set of data model components:

- Entity types (⇨ENT)
  - person
  - damage
  - policy
  - policy-holder
- Relationship types (⇨REL)
  - (none)

- Attributes (⇨ATT)
  - state
  - registration date
  - damage-field
  - damage-cause
  - third-party-insurance-type

These constructs are now to be structured according to the principles of the data model chosen. For the specification of the ERM, entity types are linked by relationship types and relationship types are assigned to at least two entity types. Furthermore, attributes are assigned to either entity or relationship types. Additional data model components resulting from these manipulations are:

- Entity types
  - (none)
- Relationship types
  - policy-holder *is-a* person
  - policy-holder *reports* damage
  - policy-holder *is_insured_by* policy
  - damage *affects* policy
  - policy *covers* damage
- Attributes (⇨ATT)
  - (none)

The ERM resulting from this assignment of data model components is depicted in Figure 4-24. This model shows that a policy-holder must have at least one policy. He may report damages which affect at least the policy of the policy-holder but can also concern policies of other policy-holders, e.g. if several cars involved in an accident are insured by policies of the company concerned.

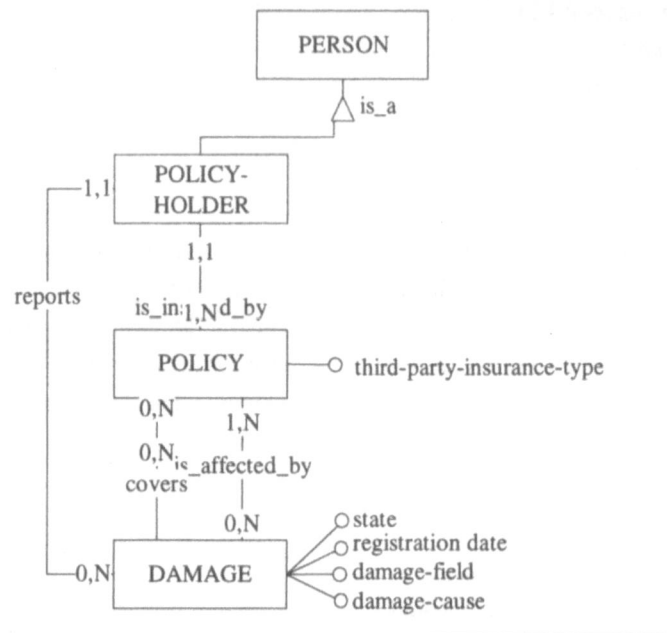

Figure 4-24: ERM for the DOM-2 example

4.2.4   Step 4: Integrity Constraints

4.2.4.1   Deliverables

The specification of a conceptual data model may result in integrity constraints which may not yet be specified as business rules. In order to have the whole description of the universe of discourse based on a single construct, i.e., business rules, it is necessary to complete the rule collection describing processes by formalizing different types of integrity constraints as business rules. The result of this modeling task is the complete conceptual data model.

4.2.4.2   Procedure

In order to find the relevant integrity constraints, the classification system discussed in Chapter 3 can be used to distinguish subtasks which concern the following constraint types[54] (cf. section 3.2.1):

- *Implicit* integrity constraints
  - *referential* integrity constraints
  - *non-referential* implicit integrity constraints

---

[54] Another procedure is discussed in Leikauf (1990), pp. 38.

- *Explicit* integrity constraints
  - *static* explicit integrity constraints
  - *dynamic* integrity constraints

The step *integrity constraint* is to be executed with respect to the following rules, which refine rule [SA-4] of the main modeling process:

Business Rule [SA-4-1] 'IMPLICIT-REFERENTIAL-CONSTRAINTS'
> ON   *(conceptual data model specified)*
> THEN   *specify referential integrity constraints as business rules;*
> *raise event 'REFERENTIAL-CONSTRAINTS-SPECIFIED'*

Business Rule [SA-4-2] 'IMPLICIT-NON-REFERENTIAL-CONSTRAINTS'
> ON   *(referential constraints specified)*
> THEN   *specify non-referential implicit integrity constraints by using*
> *business rules;*
> *raise event 'NON-REFERENTIAL-CONSTRAINTS-SPECIFIED'*

Business Rule [SA-4-3] 'EXPLICIT-STATIC-CONSTRAINTS'
> ON   *(non-referential constraints specified)*
> THEN   *specify static explicit integrity constraints by using business rules;*
> *raise event 'STATIC-CONSTRAINTS-SPECIFIED'*

Business Rule [SA-4-4] 'EXPLICIT-DYNAMIC-CONSTRAINTS'
> ON   *(static constraints specified)*
> THEN   *specify dynamic integrity constraints by using business rules;*
> *raise event 'CONSTRAINTS-SPECIFIED'*

Business Rule [SA-4-5] 'CONCLUDE-INTEGRITY-CONSTRAINTS'
> ON   *(constraints specified)*
> THEN   *verify integrity constraints;*
> *raise event '**CONCEPTUAL-MODEL-COMPLETE**'*

4.2.4.3   Application to DOM-2

At first, the conceptual data model is analyzed with respect to referential integrity constraints which results in the following set of business rules:

Business Rule REF-INT-POLICY-POLICY-HOLDER-1
> ON   *(INSERT policy) OR (UPDATE policy-holder of policy)*
> IF   *NOT (exists policy-holder related to policy)*
> THEN   *error 'Policy holder does not exist';*
> *reject operation*

Business Rule REF-INT-DAMAGE-POLICY-HOLDER
>   *ON*   *(INSERT damage) OR (UPDATE policy-holder of damage)*
>   *IF*   *NOT (exists policy-holder related to damage)*
>   *THEN*   *error 'Policy holder does not exist';*
>   *reject operation*

Business Rule REF-INT-DAMAGE-POLICY
>   *ON*   *(INSERT damage) OR (UPDATE policy-number of damage)*
>   *IF*   *NOT (exists policy related to damage)*
>   *THEN*   *error 'Policy does not exist';*
>   *reject operation*

In Figure 4-24, four attributes are depicted. The following business rules specify two of the possible *non-referential implicit integrity constraints*:

Business Rule UNIQUE-POLICY-NUMBER
>   *ON*   *(INSERT policy) OR (UPDATE policy-number of policy)*
>   *IF*   *(exists policy-number of policy)*
>   *THEN*   *error 'Policy holder does already exist';*
>   *reject operation*

Business Rule DATATYPE-REGIS-DATE
>   *ON*   *(INSERT damage) OR (UPDATE registration-date of damage)*
>   *IF*   *(registration-date of damage incorrect)*
>   *THEN*   *error 'Entered value is not a date';*
>   *reject operation*

The next types of constraints to be specified are *static, explicit integrity constraints* which may encompass the following business rules:

Business Rule REGIS-DATE-1
>   *ON*   *(INSERT damage) OR (UPDATE registration-date of damage)*
>   *IF*   *(registration-date of damage > TODAY ())*
>   *THEN*   *error 'Registration date must not be in the future';*
>   *reject operation*

Business Rule DAMAGE-STATE-1
>   *ON*   *(INSERT damage) OR (UPDATE state of damage)*
>   *IF*   *(state of damage NOT IN ('created', 'rejected',*
>   *'provisionally accepted', 'definitely accepted', 'calculated',*
>   *'closed', 'archived'))*
>   *THEN*   *error 'Illegal entry in damage-state';*
>   *reject operation*

Business Rule THIRD-PARTY-INS-TYPE-1
    ON   *(INSERT policy) OR (UPDATE third-party-insurance-type of policy)*
    IF   *(third-party-insurance-type of policy NOT IN*
            (...<list of allowed values>...))
    THEN *error 'Illegal entry in third-party-insurance-type';*
            reject operation

In order to exemplify the specification of *dynamic integrity constraints*, state transitions for the property *state* of the entity type *damage* are regarded. According to the syntax for business rules, these facts are specified as follows:

Business Rule DAMAGE-STATES-1
    ON   *(INSERT damage)*
    THEN *damage-state of damage := 'created'*

Business Rule DAMAGE-STATES-2
    ON   *(UPDATE state of damage)*
    IF   *(state of damage ('created') NOT TO ('rejected', 'provisionally-*
            accepted', 'definitively-accepted'))
    THEN *Error 'Illegal state change';*
            reject operation

Business Rule DAMAGE-STATES-3
    ON   *(UPDATE state of damage)*
    IF   *(state of damage ('provisionally-accepted') NOT TO ('rejected',*
            'definitively-accepted'))
    THEN *error 'Illegal state change';*
            reject operation

Business Rule DAMAGE-STATES-4
    ON   *(UPDATE state of damage)*
    IF   *(state of damage ('definitively-accepted') NOT TO ('closed'))*
    THEN *error 'Illegal state change';*
            reject operation

Business Rule DAMAGE-STATES-5
    ON   *(UPDATE state of damage)*
    IF   *(state of damage ('closed') NOT TO ('archived'))*
    THEN *error 'Illegal state change';*
            reject operation

Business Rule DAMAGE-STATES-6
    ON   *(DELETE damage)*
    IF   *(state of damage NOT IN ('archived', 'rejected'))*
    THEN *error 'Damage not archived or rejected; deletion not permitted';*
            reject operation

The graphical presentation of these allowed state changes can be done by the data behavior view which will be discussed in Chapter 5.

4.2.5   Step 5: Validation

The last step of *BROCOM* deals with the validation of a conceptual model[55]. Validation has the „objective of checking whether the conceptual model is consistent and whether it correctly expresses the requirements informally stated by the users"[56]. The steps can be divided in two subtasks[57]:

- The *static model validation* includes checking how well-formed the model is, i.e., „that it has been constructed according to the syntax rules of the modeling formalism"[58]. Furthermore, the consistency and completeness of the rules have to be checked, i.e., self-contradicting and inconsistent rules have to be detected.
- The *dynamic model validation* aims to prove that (1) the model's behavior is consistent, and (2) that it is consistent with respect to the user's perception of the real world.

Because the structure of this subprocess is rather trivial it is not refined by using business rules. The validation of the conceptual model may be supported by

- graphical specifications or presentations of the meta data[59],
- the abstraction levels introduced above,
- a classification of the meta data[60], or
- by simulating the behavior of the system[61].

---

[55] For a comprehensive discussion of the validation and verification of models cf. e.g. Thaller (1994).

[56] Rolland/Proix (1992), p. 259.

[57] Cf. for the following Meseguer (1990); Tsalgatidou et al. (1990), pp. 257; Sakthivel/Moily (1993).

[58] Tsalgatidou et al. (1990), p. 258.

[59] Different views on meta data are discussed in Chapter 5.

[60] May be supported by a repository system like the one described in Chapter 6.

[61] Cf. e.g. Bubenko/Wangler (1992), p. 397.

4.3 Summary

In this chapter, two main parts of **BROCOM** have been presented: the *meta model* and the *modeling tasks*. The meta model presented focuses on business rules as a main modeling construct for the specification of automation and integrity rules. Based on the meta entity types included in the meta model, the dynamic properties of business processes can be specified in a flexible way. In order to provide a comprehensive business model, additional facts are considered:

- *Processors* are separated into manual and automated processors resulting in a possible analysis and optimization of the execution of a process.
- *Organizational units* are used to define responsibilities and to embed the processors into an organizational structure to show the interdependencies between processes and organizational structures. In the context of BPR, such units would be considered as resource pools with a resource owner who may assign specific resources, i.e., also processors to the execution of business processes[62].
- Origins, i.e., people or documents which know about or contain business rules.

The conceptual modeling which is based on the meta model encompasses six main tasks of which the tasks one to four result in process structure, process description, conceptual data model, and integrity constraints.

Afterwards, the conceptual model has to be *verified* and *validated*. For these two modeling tasks, graphical models on specific views, e.g., the business processes or their interdependencies with the data structure, are very valuable. The next chapter therefore encompasses an evaluation of selected models for the representation of important views on the meta model. Though the graphical representation is often very valuable, there are limits for graphical representation, e.g., the size of a diagram. Therefore, not only graphical views but also textual analyses, i.e., reports are considered.

---

[62] Cf. e.g. Davenport (1993); Jacobson et al. (1995).

5 VIEWS ON THE META MODEL

5.1 Introduction

In the previous chapter, *BROCOM*, a new approach for conceptual modeling, was introduced which focuses on business rules as a main modeling construct. Such a conceptual model may lead to a large and thus very complex set of rules. In order to reduce this complexity and to enable the communication between all people involved in the IS development, several views on the facts are useful.

One assumption of the research project was that there should be an appropriate graphical model for each relevant view on the meta data of *BROCOM*. Thus, no completely new model is to be developed but existing ones may be modified for the *BROCOM* approach. The views that have been taken into account are

- process view,
- data view,
- view on the relationship between data and business rules/processes, and
- data object behavior view.

In this chapter, selected models for the presentation of these views are evaluated[1]; however, because there is an immense amount of graphical models, a pre-selection had to be done which was mainly based on the usability of the model for a specific view. The following models have been selected for the evaluation:

- Process view:
  - Petri net and its extensions
  - Conceptual processing model of Merise
- Data structure view:
  - Entity Relationship Model
  - NIAM
- Data usage view:
  - Data flow diagram
  - Entity-Relationship Event-Rules Model
- Data object behavior view:
  - Behavior Integrated Entity-Relationship model

---

[1] For a comparison of methodologies with respect to their capability to represent business rules cf. Herbst et al. (1994).

Other graphical models, e.g., for the representation of views on processors and organizational units, are not discussed.

The graphical models are applied to the example business rules from case study DOM-2 which have been introduced in Section 3.1.3.

5.2 Process View

Business rules are dynamically linked by events resulting in ECA[2] chains which, depending on the precision attempted by the analyst, may be very precise. The *graphical presentation* of the process view has two main goals: it should be

- exact and thus unambiguous and
- understandable and readable for non-specialists, e.g., end users.

Because these goals are contradictory, the application of two complementing graphical presentations of the process views may be appropriate. One of the models should emphasize the precision and the second the readability. In the following, Petri net models and the conceptual processing model are discussed and applied to the example rules[2].

5.2.1 Petri Net

5.2.1.1 Introduction

Petri nets are formal models to specify control flows in a system[3]. They provide mechanisms for describing systems that may exhibit asynchronous and concurrent activities. A Petri net may be described by a four-tuple $N = <P,T,I,O>$, where P and T are sets of places and transitions, respectively, I is the input function which defines the set of input places of each transition and O is the output function which defines the set of output places of each transition[4]. Places in Petri nets represent states of a system. Transitions are either events or actions which lead to other states[5]. Figure 5-1 gives an example of a Petri net and its application to a job execution process.

---

[2] A comprehensive discussion of models for specifying and visualizing behavioral requirements is given in Davis (1990), pp. 21.

[3] For the basic theory on Petri nets cf. Petri (1962). Applications of Petri nets to system modelling are discussed e.g. in Genrich/Lauterbach (1981); Rosenstengel/Winand (1982); Nazareth (1993); Keen/Lakos (1994); Lee et al. (1994); Richter (1984a); Richter (1984b); Tsalgatidou et al. (1994).

[4] Cf. Sølvberg/Kung (1993), pp. 500.

[5] Cf. Rosenstengel/Winand (1982), pp. 8 Another alternative are event/condition nets which are discussed e.g. in Rosenstengel/Winand (1982), p. 66.

In a marked Petri net, as it is assumed in this work, tokens are used to describe and simulate the behavior of systems. The tokens move from place to place due to firing transitions according to the following rule: a transition is said to be enabled and fires if each of its input places contains at least one token. The application of the firing rule leads to the basic situations put together in Figure 5-2.

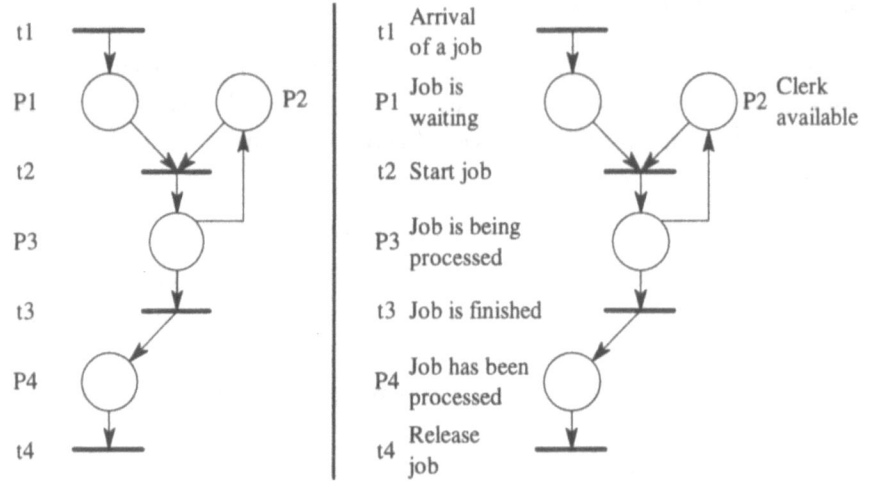

Figure 5-1:   A Petri net and its possible interpretation[6]

6   Sølvberg/Kung (1993), p. 500.

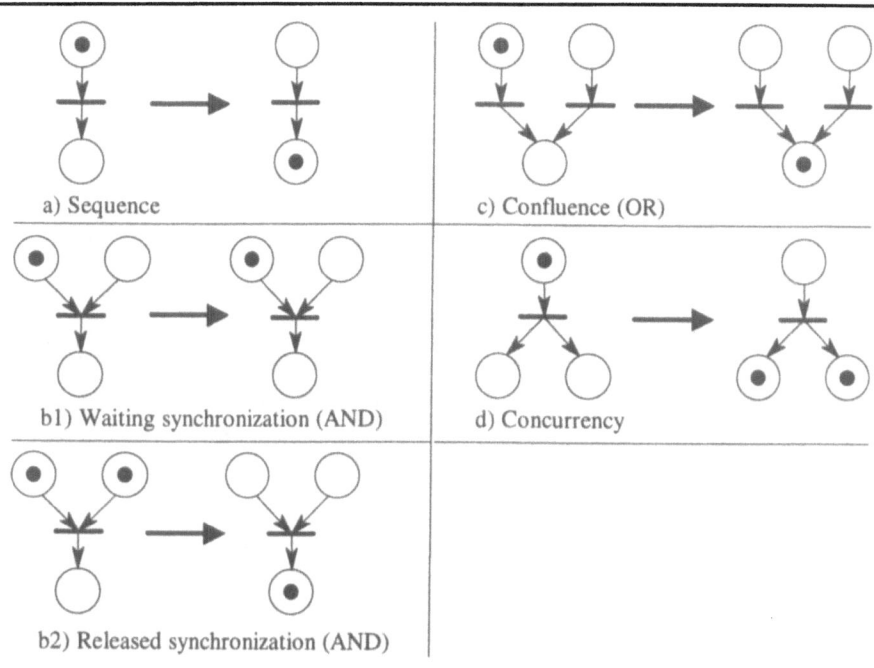

Figure 5-2: Basic situations of token flows

In order to model conditional branching and different types of complex events (e.g., temporal dependencies), the basic constructs depicted in Figure 5-2 have to be complemented. For the specification of conditions the extension depicted in Figure 5-3 is needed.

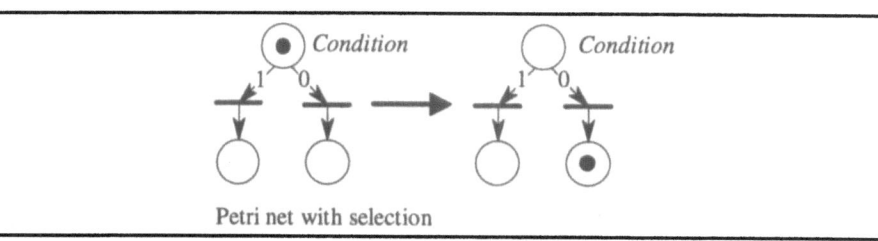

Figure 5-3: Token flow in case of a condition evaluated to false

5.2.1.2   Application

5.2.1.2.1   ECA^2 Net

As mentioned above, places of Petri nets represent states of a system, and transitions symbolize events or actions leading to other states. In an application of the Petri net to ECA rules, Tanaka[8] assigns events to places and the condition and action component of a business rule to transitions (cf. Figure 5-4): an E/R network is a high-level Petri Net defined by the triple (E,R,A): E is a finite set of places representing events, R is a finite set of transitions which represent rules (condition/action part of a business rule) and A is a finite set of directed arcs denoting the raising of events and the triggering of rules.[9] To model business rules structured as ECAA, this assignment is modified in such a way that places repre-

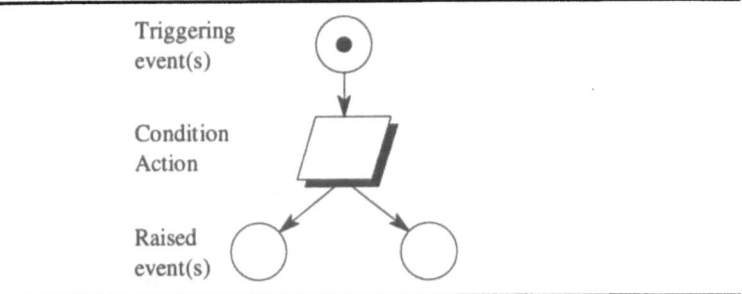

Figure 5-4:   Petri net for an ECA rule[7]

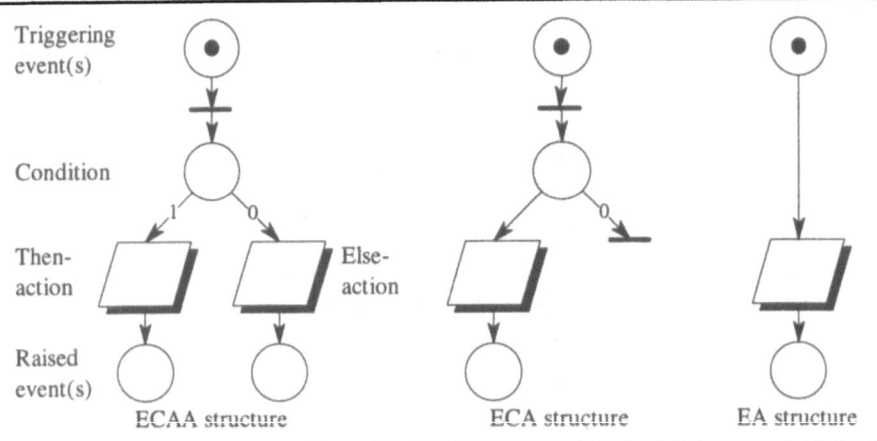

Figure 5-5:   Petri nets for a business rule structured as ECAA, ECA, or EA

---

[7]   Adapted from Tanaka (1992), p. 69.
[8]   Cf. Tanaka (1992), pp. 104.

sent either events or conditions and transitions are either dummies or represent the action component (cf. Figure 5-5). The ECA^2 net resulting from the application of these symbols is also a high-level Petri net which is intended to exactly represent the behavior of a system and can be used as a basis for the simulation of processes.

In Figure 5-5, a triggering event only consists of a single place; however, in order to represent the semantics of composite events, a more complex representation is needed (cf. Figure 5-6). This involves several places representing the constituting events and the exact logic of composite events as depicted in Figure 5-7 and Figure 5-8[10]. Between all events involved, dummy transitions have to be drawn because directed arcs between places are not allowed in a Petri net.

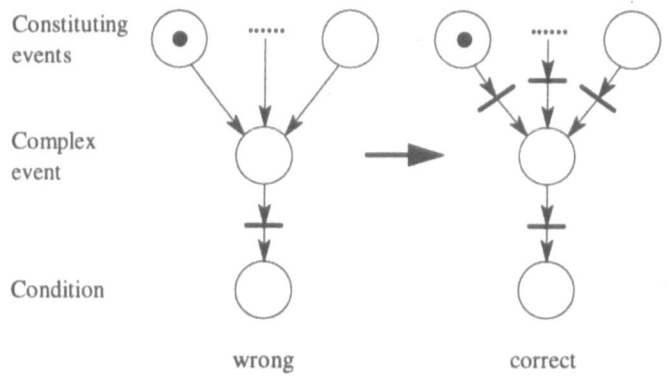

Figure 5-6: Dummy transitions for complex events[11]

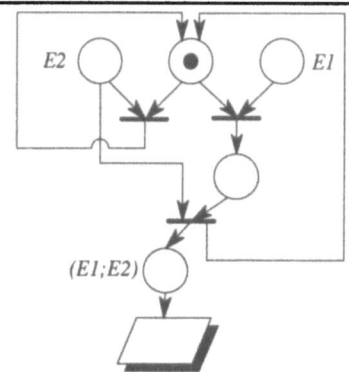

Figure 5-7: Petri net for the sequence event {E1;E2}[12]

---

9   Cf. Tanaka (1992), pp. 116.
10 Other examples of complex events modeled by Petri nets are discussed in Gatziu (1994).
11 Cf. Tanaka (1992), p. 69.

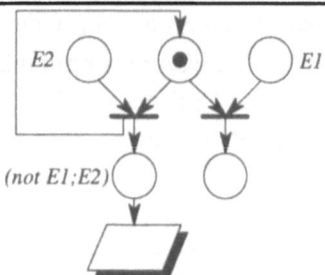

Figure 5-8: Petri net for the sequence event {NOT E1; E2}[13]

An event may trigger several business rules and may be relevant for several complex events. Thus, a token has to be multiplied, which can be achieved by a concurrency as already depicted in Figure 5-2d. The correct representation of the concurrency of events is given in Figure 5-9.

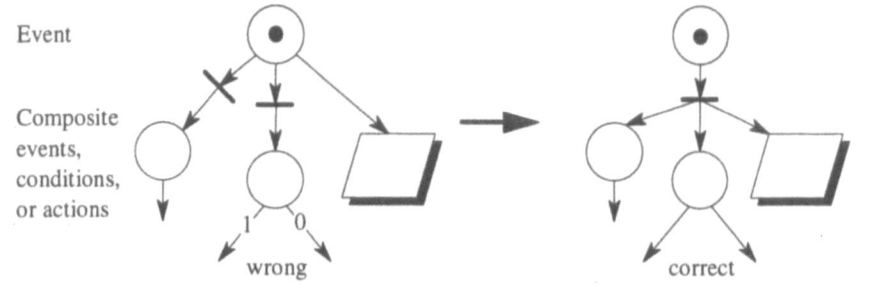

Figure 5-9: Concurrency in rule triggering

The application of this notation to the example business rules leads to the ECA^2 net depicted in Figure 5-10.

---

[12] Gatziu (1994), p. 102.
[13] Gatziu (1994), p. 112.

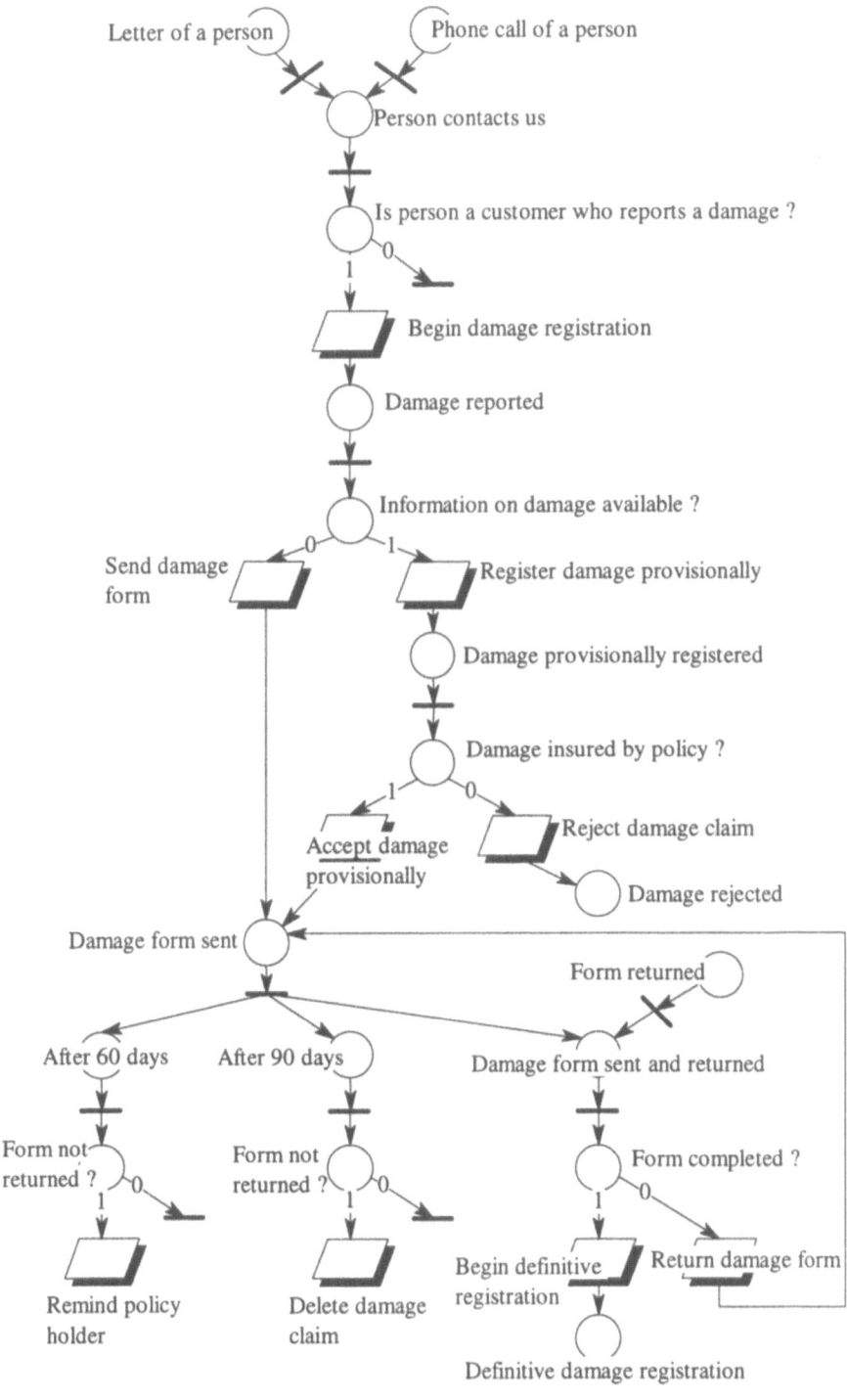

Figure 5-10: ECA² net for the DOM-2 example

5.2.1.2.2 Simplified ECA^2 Net

The ECA^2 net provides a very precise representation of a process and is an appropriate basis for e.g. simulations. However, in order to communicate about a process, a reduction of the complexity may be necessary. Therefore, a simplified ECA^2 net is introduced which is easier to understand, but may lack of precision. The simplification is achieved

- by hiding details of the event structures (cf. Figure 5-11) and
- by depicting the if-then-else structure in a single transition (cf. Figure 5-12).

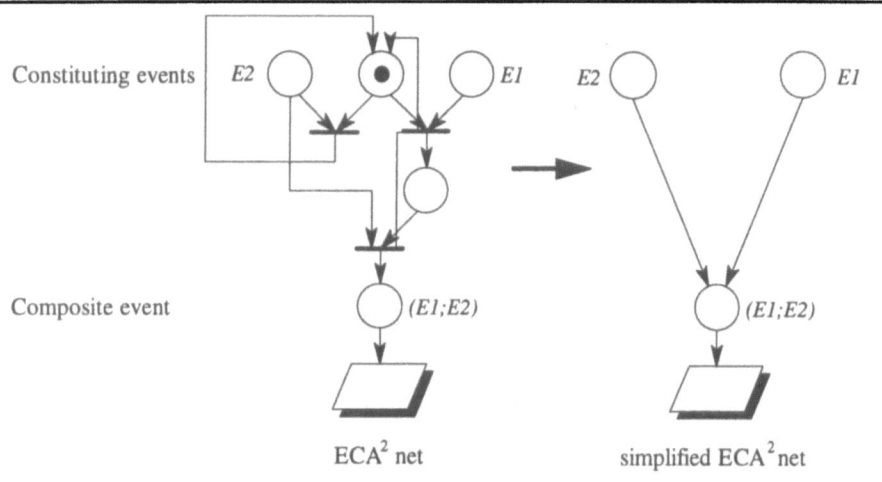

Figure 5-11: Composite event structure as a simple ECA^2 net

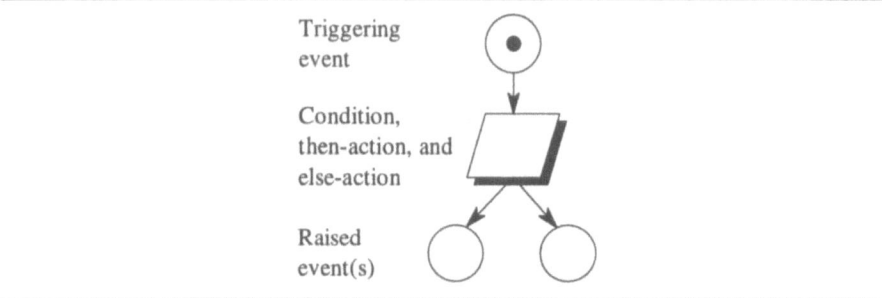

Figure 5-12: Business rule structure in a simplified ECA^2 net

Applying these modifications to the ECA^2 net of Figure 5-10 results in a simplified representation of the process (cf. Figure 5-13). However, because of the adaptations, this resulting net is no longer a Petri net because it violates basic properties of Petri nets (e.g., by directly connecting places in composite events).

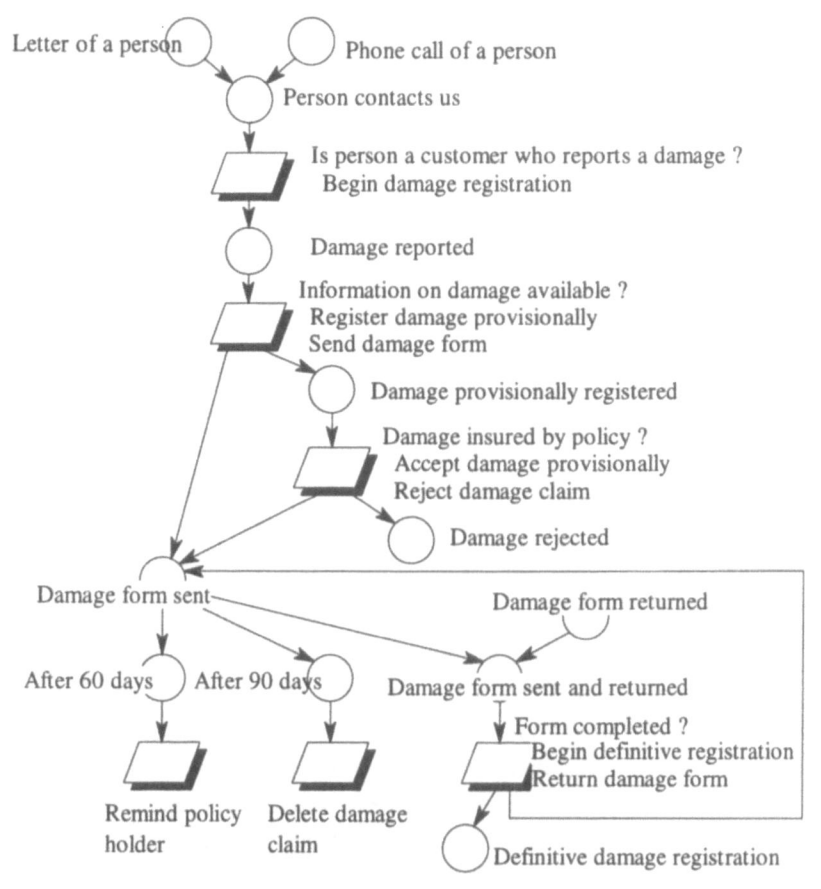

Figure 5-13: Simplified ECA^2 net for the DOM-2 example

5.2.2 Conceptual Processing Model

5.2.2.1 Introduction

The methodology Merise uses the conceptual processing model (CPM) for the specification of processes. This model encompasses triggering events, (synchronized) operations, and resulting events[14] (cf. Figure 5-14). An operation consists of one or more tasks that are based on management rules and are executed sequentially. Every operation may lead to different events according to issuing rules which may e.g. be 'operation has been successful' or 'operation has

---

[14] Quang/Chartier-Kastler (1991), pp. 46. An early version of CPM is discussed in Rochfeld/Tardieu (1983), pp. 238. An overview of Merise is given in Gray/Rao (1993), pp. 86.

failed'; the abbreviation NR signifies that *no response* will follow. In CPM an event is seen as the carrier of properties; these correspond with attributes in the Merise data model. Merise/2 introduces object life-cycles in which the events changing the states of a certain object are modeled[15].

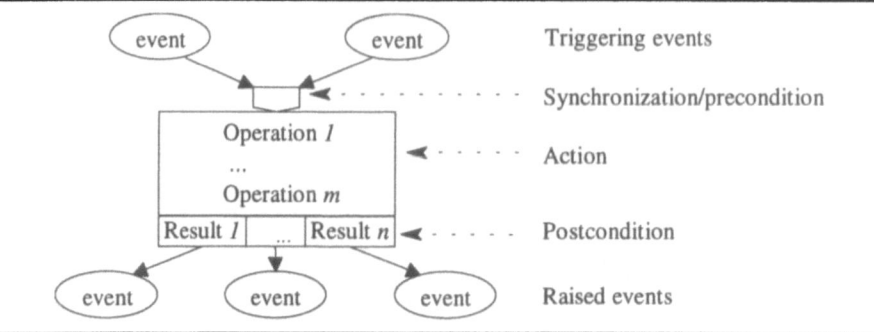

Figure 5-14: Symbols used in a CPM[16]

5.2.2.2   Application

The application of CPM to the presentation of a process consisting of business rules requires some adaptations of the semantics. Elementary events can be directly assigned to the event construct of CPM. However, composite events (e.g., *Person contacts us* as a disjunction of *Phone call of a person* and *Letter of a person*) are part of the synchronization element, whereas the specification of the composite event is done verbally. Conditions of business rules are also assigned to the synchronization thus leading to an ambiguity of this construct. The action is finally represented by the operation construct.

Using these semantic assignments, the example rules can be shown by the CPM depicted in Figure 5-15[17]. However, the application of CPM to the representation of processes specified by business rules results in the following problems:

- The precise logic of composite events cannot be depicted, but only described verbally. This leads to a lower precision of the process representation similar to the simplified ECA^2 net.
- Composite events and conditions are put together in a single construct.
- CPM does not support a selection between actions because the precondition only decides on the execution of the following operations. Therefore, selections lead to redundant conditions which have to be specified once as a negation, e.g., *Information available* and *NOT (Information available)*.

---

[15] Tardieu (1992).

[16] Cf. Tardieu (1992), pp. 50.

[17] Another application of CPM to business rules is given in Herbst et al. (1994), p. 34.

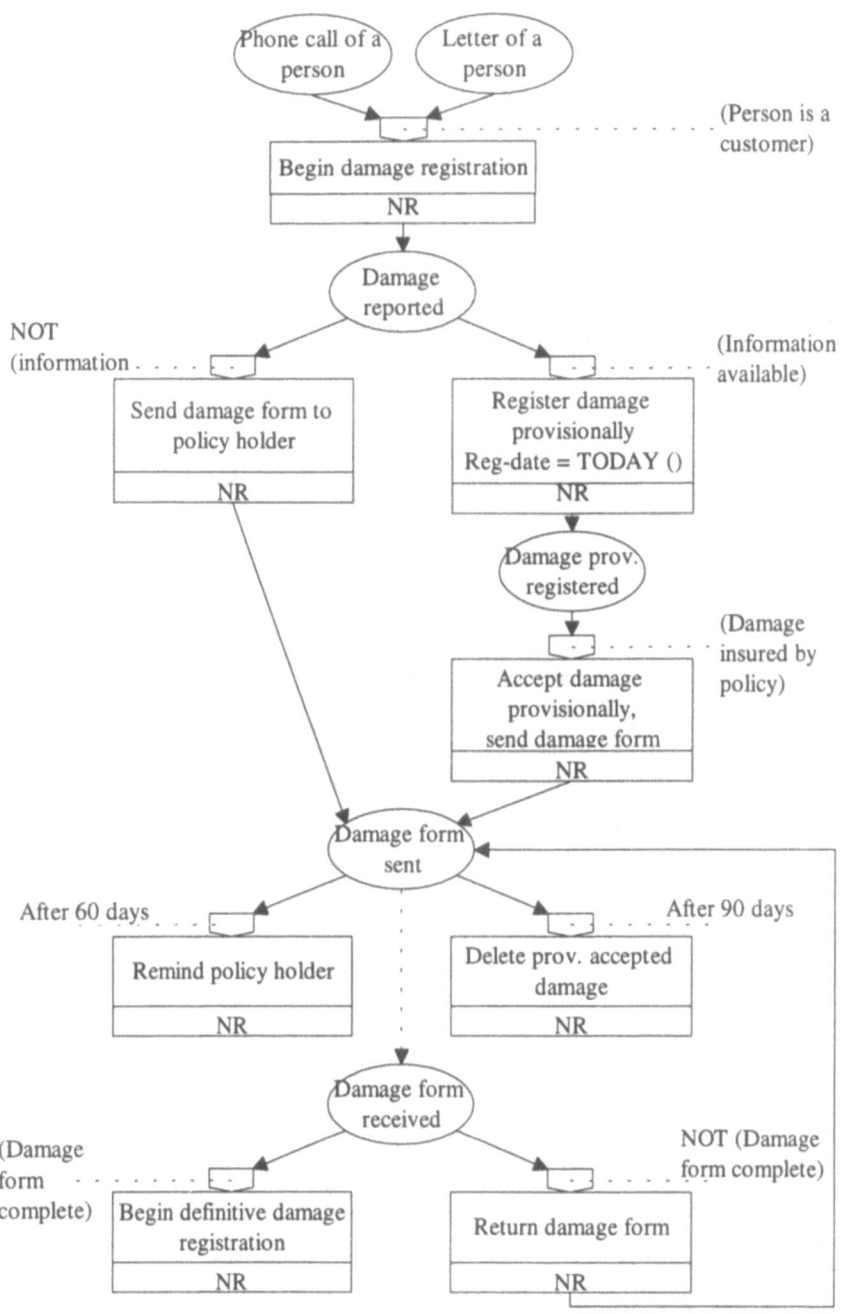

Figure 5-15: Conceptual processing model for the DOM-2 example

Though the CPM may be applied to show the process view of the meta model, the combination of the ECA^2 net and the simplified ECA^2 net is considered as the better alternative for the presentation of processes specified by business rules. Another argument in favor of the Petri net is the availability of tools based on them (e.g. for simulating the behavior of a process).

5.3   View on the Data Model

The data structure view includes a graphical representation of the facts described in the submodel *Data model component*. The meta entity types included in the meta model support the specification of an Entity-Relationship model (ERM). In addition, because of its expressiveness and maturity, NIAM is discussed. Object-oriented data models are not considered because they are closely related to the ERM[18].

5.3.1   Entity-Relationship Model

5.3.1.1   Introduction

Since its introduction[19], the ERM has become the most popular model for conceptual database design[20] and is still a major topic of international conferences. It has furthermore been chosen for the IRDS standard of ANSI[21].

The ERM encompasses the basic constructs entities, relationships, and attributes. In literature the terms entity and entity type are used heterogeneously. Some authors use the term *entity* for a class of real-world objects[22] and others for the objects themselves, referring to the classes as *entity types*. In the following the term *entity* is used for unique and identifiable objects and the term *entity type* for sets of similar entities. Accordingly, the term *relationship* is used for specific relationships between entities and the term *relationship types* for sets of similar relationships.

In an ERM, entity types are represented by a rectangle. The properties common to all entities of a specific type are called attributes and are depicted as circles or ellipses. Relationships mutually associate at least two entities, resulting in binary or n-ary relationships. Each association of entities in a relationship has a cardi-

---

[18]   Chen (1992), p. 1.

[19]   Cf. Chen (1975); Chen (1976).

[20]   Lockemann/Radermacher (1990), pp. 8.

[21]   Cf. ANSI (1989).

[22]   Cf. Batini et al. (1992), p. 31.

nality which restricts the possible minimum and/or maximum of instances of the related entity type(s)[23].

There are several extensions of the ERM[24] introducing symbols to specify additional semantics like *is-a* relationship types for specialization[25].

5.3.1.2 Application

An ERM is a static view on the universe of discourse and thus does not allow for the explicit formulation of events or actions[26]. The business rules embedded in an ERM are often implicit integrity constraints which are expressed in cardinalities resulting in e.g. existence dependencies. However, an ERM may imply events which can be derived by determining all data manipulations that may violate the constraint:

> *Example: Existence dependency* Customer ⇨ Order
> *The potentially violating events would be the insertion, the modification, and the deletion of the relationship between a customer and an order.*

The condition part of a business rule is represented by certain implicit integrity constraints[27] like the existence dependency used in the example above.

Actions to be executed in case of a constraint violation cannot be expressed in or implied from the ERM without additional natural language descriptions which are neglected in this context[28]. For the example above, the following actions can be derived[29]:

> *Example: Existence dependency* Customer ⇨ Order
> *The reaction on a violation can be either a rejection or a cascading correction.*

As discussed in Chapter 4, business rules encompass references to components of the data model. These references can be used to automatically derive e.g. entity types from a set of business rules. To support this derivation of different data model components, the three keywords ⇨ENT, ⇨REL, and ⇨ATT are used

---

[23] In literature, there exist a large amount of different notation to express cardinalities. For an overview cf. Lenzerini/Santucci (1983); Knolmayer/Myrach (1990), p. 93; Ferg (1991), pp. 1.

[24] Cf. Sinz (1990), pp. 17.

[25] Cf. Smith/Smith (1977); Batini et al. (1992), pp. 35.

[26] Cf. Herbst et al. (1994).

[27] For implicit integrity constraints cf. Chapter 3.

[28] For such verbal specifications cf. e.g. Jarke/Pohl (1992), pp. 355.

[29] A systematic evaluation of possible actions is given e.g. in Lazarevic/Misic (1991), pp. 98.

within the rule formalization. The application of these keywords within the rule formalization leads to the following rules[30]:

Business rule [1] 'PERSON-CONTACTS-US'
    ON   *(PHONE-CALL-OF-PERSON) OR (LETTER-OF-PERSON)*
    IF   (⇨ENT person ⇨REL is-a ⇨ENT policy-holder) AND*
           (⇨ENT policy-holder ⇨REL reports ⇨ENT damage)
    THEN *begin damage-registration;*
           raise event 'DAMAGE-REPORTED'

Business rule [2] 'REGISTER-PROV-DAMAGE'
    ON   *(⇨ENT damage reported)*
    IF   *(information about ⇨ENT damage available)*
    THEN *insert ⇨ENT damage;*
           ⇨ATT registration-date := TODAY ();
           ⇨ATT state := 'provisional';
           raise event 'DAMAGE-PROV-REGISTERED'
    ELSE   *send ⇨ENT damage form to ⇨ENT policy-holder;*
           raise event 'DAMAGE-FORM-SENT'

Business rule [3] 'ACCEPT-PROV-DAMAGE'
    ON   *(provisional ⇨ENT damage registered)*
    IF   *(⇨ENT Damage ⇨REL is_covered_by ⇨ENT policy)*
    THEN *accept ⇨ENT damage provisionally;*
           send damage form to ⇨ENT policy-holder;
           raise event 'DAMAGE-FORM-SENT';
           raise event 'DAMAGE-PROV-ACCEPTED'
    ELSE   *reject ⇨ENT damage;*
           raise event 'DAMAGE-REJECTED'

Business rule [4] 'REMINDER-FOR-DAMAGE-FORM'
    ON   *(60 days after (DAMAGE-FORM-SENT))*
    IF   *(damage form not returned)*
    THEN *remind policy-holder*

Business rule [5] 'TIME-OUT-FOR-DAMAGE-FORM'
    ON   *(90 days after (DAMAGE-FORM-SENT))*
    IF   *(damage form not returned)*
    THEN *delete provisionally accepted ⇨ENT damage*

---

[30] The use of the keywords of course reduces the readability of the rules; however, in a report of registered rules the keywords could be omitted.

Business rule [6] 'DEF-REGISTER-DAMAGE'
> ON   *(DAMAGE-FORM-RECEIVED WITHIN (90 days after*
>   *(DAMAGE-FORM-SENT)))*
> IF   *(information is complete)*
> THEN   *raise event 'START-DEF-DAMAGE-REGISTRATION'*
> ELSE   *return form to policy-holder;*
>   *raise event 'DAMAGE-FORM-SENT'*

Business rule [7] 'DAMAGE-COVERED'
> ON   *((⇨ATT damage-field entered) OR (⇨ATT damage-cause entered))*
> IF   *(⇨ATT damage-field = private third party insurance) AND*
>   *(⇨ATT damage-cause = damage of a car in use) AND*
>   *(⇨ATT third-party-insurance-type = family, single or senior)*
> THEN   *issue error message „Damages of cars in use are not insured by this*
>   *policy;*
>   *please check and if necessary, pass the file to the central office. "*

From these business rules the following set of constructs for an ERM may be derived:

- Entity types (⇨ENT):
  - person
  - damage
  - policy
  - policy-holder
- Relationship types (⇨REL)
  - policy-holder *is-a* person
  - policy-holder *reports* damage
  - damage *is_covered_by* policy
- Attributes (⇨ATT)
  - state
  - registration date
  - damage-field
  - damage-cause
  - third-party-insurance-type

For the specification of the ERM, entity types are linked by relationship types, relationship types have to be assigned to at least two entity types, and attributes to entity or relationship types. This may lead to additional constructs completing the list given above. The list of data model components encompasses additionally

- Entity types
  - (none)

- Relationship types
  - policy-holder *is_insured_by* policy
  - damage *affects* policy
- Attributes
  - (none)

Finally, these modeling constructs may be depicted in an ERM (cf. Figure 5-16).

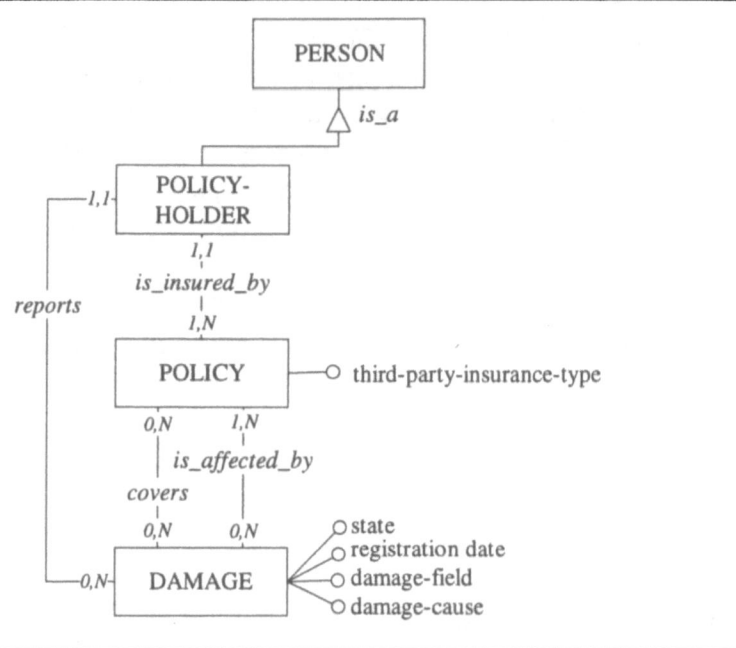

Figure 5-16: ERM for the DOM-2 example

The implicit integrity constraints represented by this ERM are among others

- Is-a relationship between *PERSON* and *POLICY-HOLDER,*
- existence dependency of *POLICY* and *POLICY-HOLDER*, and
- existence dependency of *DAMAGE* and *POLICY-HOLDER.*

As described in Chapter 4, these constraints additionally have to be specified as business rules in order to make them explicit and unambiguous.

The graphical presentation of the ERM may support the validation of business rules which is done in the fifth step of the *BROCOM* approach. Therefore, all business rules have to be retrieved which are classified as integrity rules and which refer a specific entity type from within their condition. The selection of all integrity rules referring the entity type *policy* could result in a list which may encompass the following rule:

> ON   (insert policy) OR (update policy-holder of policy)
> IF   NOT (exist policy-holder related to policy)
> THEN   reject operation

In order to verify the verbal specifications, these business rules have to be manually compared with the conceptual data model.

5.3.2 NIAM

5.3.2.1 Introduction

Another well-known data modeling technique is NIAM[31] (also known as binary semantic model[32]) which was first published in the 70's[33]. In their approach for a schema conversion between the extended ERM and NIAM Song and Forbes state that „semantically, the extended ERM is a proper subset of NIAM"[34]. Because NIAM is generally not as well known as the ERM, the symbols are explained in detail and illustrated by an example.

NIAM is a fact-oriented approach[35] and consists of two constructs for structure representation: object types and n-ary role associations (facts) where usually n > 1. Object types are further distinguished into lexical object types (LOT) and non-lexical object types (NOLOT). Fact types connected with NOLOT's are called *bridge types* (similar to attributes in the ERM) and those connecting two LOT's are called *idea types* (similar to relationships in the ERM). The graphical symbols used in NIAM diagrams are put together in Figure 5-17.

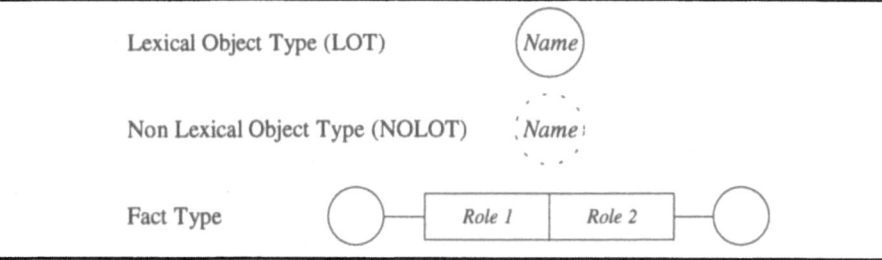

Figure 5-17: Symbols used in NIAM

To express integrity constraints, NIAM provides the constructs uniqueness and mandatoriness which are equivalent to the cardinality constraints of the ERM.

---

[31] NIAM is an acronym for Nijssen's Information Analysis Methodology.
[32] Cf. e.g. Kent (1983); Mark (1983).
[33] Cf. Falkenberg (1976).
[34] Song/Forbes (1991), p. 418.
[35] Cf. Nijssen/Halpin (1989), p. 31.

The uniqueness of roles is expressed in NIAM diagrams using arrows above the role(s) of a fact type (cf. Figure 5-18)[36].

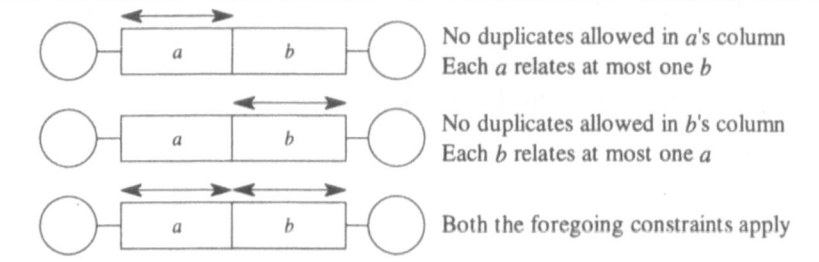

	No duplicates allowed in *a*'s column Each *a* relates at most one *b*
	No duplicates allowed in *b*'s column Each *b* relates at most one *a*
	Both the foregoing constraints apply

Figure 5-18: Uniqueness constraints in NIAM[37]

The uniqueness constraints only restrict the maximum of instances participating in a fact type. Additionally, mandatory and optional roles are depicted using either a dot or a ⊻ (cf. Figure 5-19)[38].

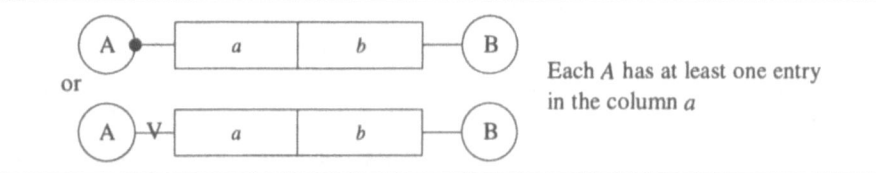

Each *A* has at least one entry in the column *a*

Figure 5-19: Mandatoriness constraints in NIAM

Furthermore, NIAM supports additional constraints between object types, such as uniqueness between different roles, equality, and exclusion[39].

The example diagram in Figure 5-20 depicts the two LOTs *customer* and *order* which are connected by a fact type. The roles of this idea type are read as Customer *puts* order and Order *belongs to* customer, respectively. The uniqueness constraint for the role *belongs to* signifies, that an order belongs to at most one customer; moreover, because the role *belongs to* is mandatory (represented by the dot at the LOT *order*), an order belongs to exactly one customer. The role *puts* on the other side has no uniqueness constraint, but also a mandatoriness expressed by the dot at the LOT *customer*. Therefore, a customer has to put at least one order. The bridgetype with the role *Reg-date* represents an attribute of *order* and is of the type *Date*.

---

[36] Cf. Nijssen/Halpin (1989), pp. 66.
[37] Nijssen/Halpin (1989), p. 75.
[38] Cf. Nijssen/Halpin (1989), pp. 114.
[39] Cf. Nijssen/Halpin (1989), pp. 171.

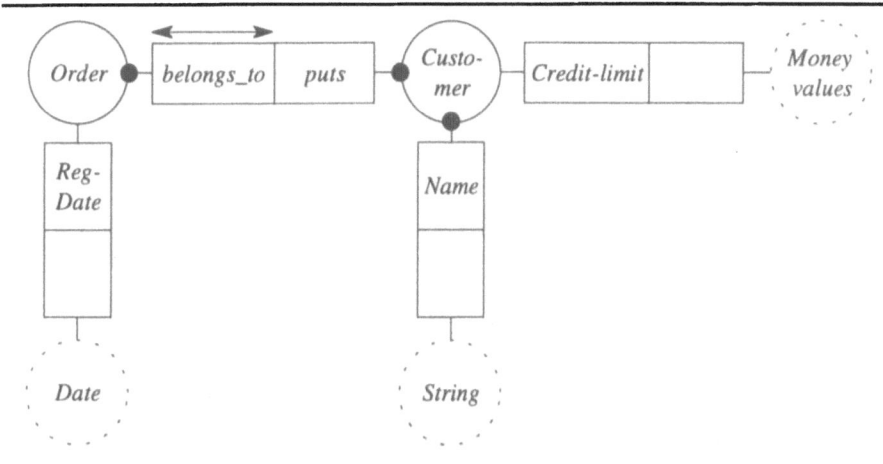

Figure 5-20: Example NIAM data model

In the context of a business rule-oriented conceptual modeling, NIAM can be used in two different ways. First, the data model can be expressed by the terms of the extended ERM and afterwards for the presentation be transformed into NIAM[40]. Second, the data model may be directly expressed in the terms of NIAM; however, this requires an adaptation of the meta model (cf. Figure 5-21).

---

[40] Rules for such a transformation can be found in Song/Forbes (1991).

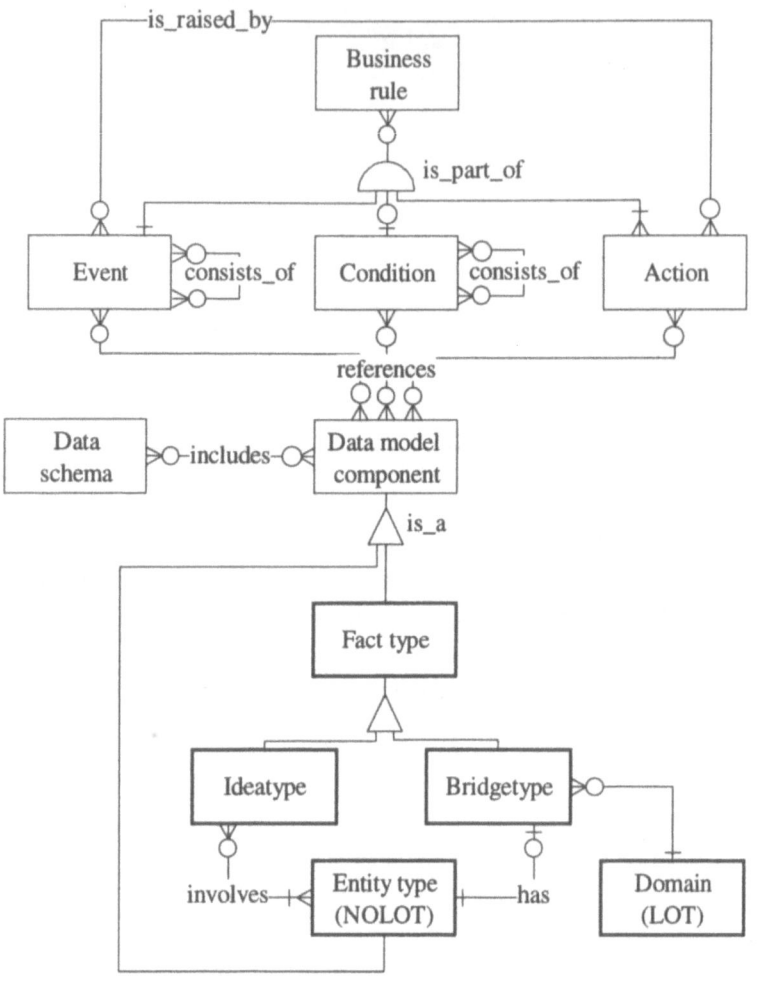

Figure 5-21: Adaptation of the meta model for using NIAM as data model

5.3.2.2   Application

For the specification of a conceptual data model expressed by the terms of NIAM, different keywords in the formalization of business rules are needed, one for fact types (⇨FACT) and one for object types (⇨OBJ). Because of the fact-orientation of NIAM, the example rules already discussed for the ERM are specified differently:

Business rule [1] 'PERSON-CONTACTS-US'
    ON   *((PHONE-CALL-OF-PERSON) OR (LETTER-OF-PERSON))*
    IF   (⇨FACT person_is-a_policy-holder) AND*
            (⇨FACT policy-holder_reports_damage)
    THEN *begin damage registration;*
            raise event 'DAMAGE-REPORTED'

Business rule [2] 'REGISTER-PROV-DAMAGE'
    ON   *(damage reported)*
    IF   *(information about ⇨OBJ damage available)*
    THEN *insert ⇨OBJ damage;*
            ⇨FACT state := 'provisional';
            ⇨FACT registration-date := TODAY ();
            raise event 'DAMAGE-PROV-REGISTERED'
    ELSE *send damage form to ⇨OBJ policy-holder;*
            raise event 'DAMAGE-FORM-SENT'

Business rule [3] 'ACCEPT-PROV-DAMAGE'
    ON   *(provisional damage registered)*
    IF   (⇨FACT damage_covered_by_policy)*
    THEN *accept ⇨OBJ damage provisionally;*
            raise event 'DAMAGE-PROV-ACCEPTED';
            send damage form to ⇨OBJ policy-holder;
            raise event 'DAMAGE-FORM-SENT'
    ELSE *reject ⇨OBJ damage;*
            raise event 'DAMAGE-REJECTED'

Business rule [4] 'REMINDER-FOR-DAMAGE-FORM'
    ON   *(60 days after (DAMAGE-FORM-SENT))*
    IF   *(damage form not returned)*
    THEN *remind ⇨OBJ policy-holder*

Business rule [5] 'TIME-OUT-FOR-DAMAGE-FORM'
    ON   *(90 days after (DAMAGE-FORM-SENT))*
    IF   *(damage form not returned)*
    THEN *delete provisionally accepted ⇨OBJ damage*

Business rule [6] 'DEF-REGISTER-DAMAGE'
    ON   *(DAMAGE-FORM-RECEIVED WITHIN (90 days after*
            (DAMAGE-FORM-SENT)))
    IF   *(information is complete)*
    THEN *raise event 'START-DEF-DAMAGE-REGISTRATION'*
    ELSE *return form to ⇨OBJ policy-holder;*
            raise event 'DAMAGE-FORM-SENT'

Business rule [7] 'DAMAGE-COVERED'

    ON   *(⇨FACT damage-field entered) OR (⇨FACT damage-cause entered)*

    IF   *(⇨FACT damage-field = private third party insurance) AND*

            (⇨FACT damage-cause = damage of a car in use) AND

            (⇨FACT third-party-insurance-type = family, single or senior)

    THEN *issue error message „Damages of cars in use are not insured by this policy;*

            please check and if necessary, pass the file to the central office. "

Using this fact-oriented formalization of the example rules, the following modeling constructs can be derived:

- Facts types (⇨FACT; either ideatype or bridgetype)
  - person_is-a_policy-holder
  - policy-holder_reports_damage
  - damage_covered_by_policy
  - state
  - registration-date
  - damage-field
  - damage-cause
  - third-party-insurance-type

- Non-lexical object types (⇨OBJ)
  - damage
  - policy-holder

For the specification of a NIAM data model, this list of modeling constructs has to be structured:

- Fact types representing idea types:
  - person_is-a_policy-holder (NOLOT's: *person* and *policy-holder*)
  - policy-holder_reports_damage (NOLOT's: *policy-holder* and *damage*)
  - damage_covered_by_policy (NOLOT's: *damage* and *policy*)

- Fact types representing bridge types:
  - state (NOLOT: *damage*)
  - registration-date (NOLOT: *damage*)
  - damage-field (NOLOT: *damage*)
  - damage-cause (NOLOT: *damage*)
  - third-party-insurance-type (NOLOT: *policy*)

- Non-lexical object types:
  - damage
  - policy
  - person
  - policy-holder

- Lexical object types:
  - damage states
  - damage fields
  - date
  - damage causes
  - insurance types

The use of these constructs in a NIAM diagram is depicted in Figure 5-22.

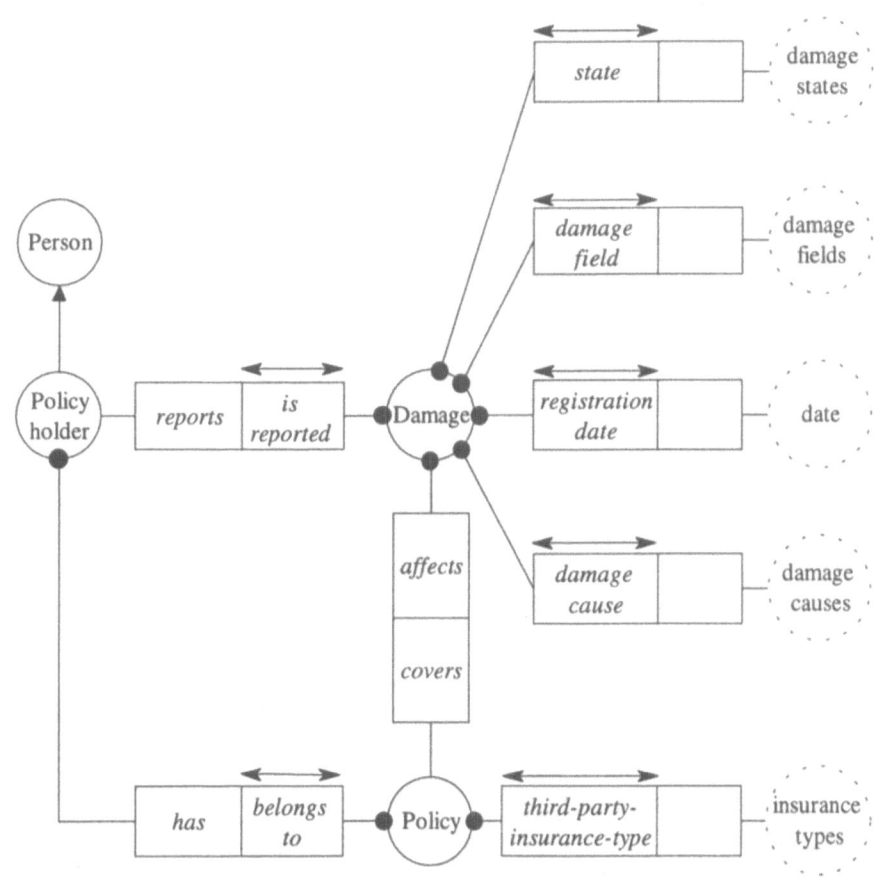

Figure 5-22: NIAM diagram for the DOM-2 example

NIAM as well as the ERM may be complemented by specification languages[41] which can be used to describe additional semantics. However, in the *BROCOM* approach, such an extension would enlarge the redundancy between data model (including the specification language) and verbally specified business rules which

---

[41] Loucopoulos et al. (1990); Theodoulidis et al. (1994); De Troyer et al. (1988).

leads to a more complex consistency check; therefore, these specification languages are deliberately omitted.

This section on NIAM presumed the use of this adapted meta model for the conceptual modeling; however, the following sections are again based on the *BROCOM* meta model introduced in Chapter 4.

5.4   View on the Relationship between Data Structure and Processes

The view on the data model discussed in the preceding section depicts the static data structure of the conceptual model. The use of data model components in business rules, i.e., also in processes described by rules, is not represented. Thus a view which explains these relationships is needed. Two models have been preselected for this purpose. The first is the well known and widely used Data Flow Diagram (DFD) and the second the rather new Entity-Relationship Event-Rule $(ER)^2$ model[42].

5.4.1   Data Flow Diagrams

5.4.1.1   Introduction

Data flow diagrams are an important technique of Structured Analysis[43] (and related methodologies) and are used to partition a system. A DFD does not explicitly represent events or conditions but only active components of the system and the data interfaces between them. In the specification of a DFD four different symbols are used (cf. Figure 5-23)[44]:

- *Data flow*: Portrays a data path between processes or between a process and either an external unit or a data store.
- *Process*: Transforms the structure or the content of data.
- *Data store*: Represents a file or data base.
- *Source/Sink*: Stands for a net originator or receiver of data and is typically a person or organization outside the system.

---

[42] Of course, also a non-graphical representation like an adapted entity/process matrix as introduced by Martin (1990), pp. 269 could be applied.

[43] Cf. De Marco (1978); Gane/Sarson (1979); McMenamin/Palmer (1984).

[44] Cf. De Marco (1978), p. 40.

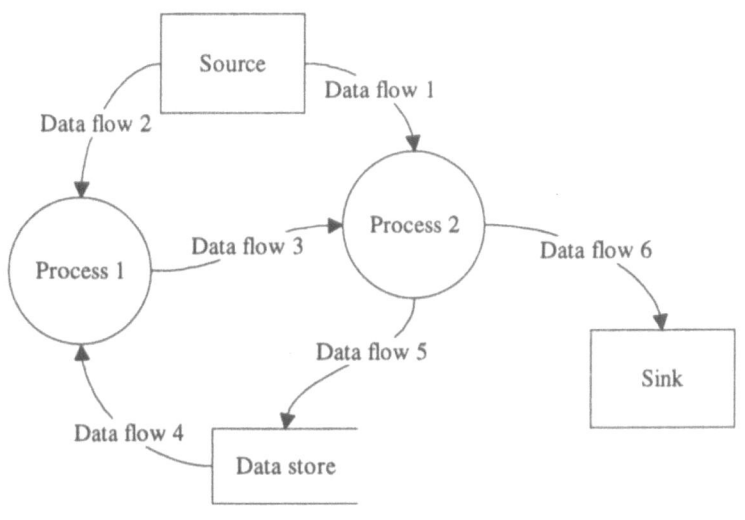

Figure 5-23: Symbols used in a DFD[45]

It is important to mention that a DFD does not imply any sequences of process execution, i.e., the arrival of a data flow does not necessarily lead to the execution of a process. De Marco states that „the Data Flow Diagram portrays a situation from the point of view of data, while a flowchart portrays it from the point of view of those who act upon the data. (...) Since no control is shown, you can't tell from looking at a Data Flow Diagram which path will be followed. The Data Flow Diagram shows only the set of possible paths. Similarly, you can't tell what initiates a given process."[46] Data flows do sometimes but not always coincide with events: there are data flows that do not directly result from the occurrence of an event and there are time and control events that cannot be represented as data flows[47]. Therefore, the semantics represented by business rules cannot unambiguously be assigned to symbols of DFD's.

5.4.1.2   Application

The applicability of the DFD as a graphical presentation of business rules is rather limited[48]. From the diagram depicted in Figure 5-24 it becomes obvious that DFD do not adequately represent the semantics of business rules and their

---

[45] Page-Jones (1990), p. 151.

[46] De Marco (1978), p. 40.

[47] Yourdon (1989), p. 340.; In Robson/Henderson (1993) the events and event stores are explicitly visualized in the DFD; however, the precise process sequences are still not represented.

[48] Cf. Herbst et al. (1994), pp. 33.

interdependencies with the data model. In order to at least express the initiation of sending the reminder (business rule 4) or the deletion of a provisionally registered damage (business rule 5), a source *clock* has been introduced which sends a temporal 'dataflow' to the processes *Send reminder* and *Delete provisional damage*. However, the fact that the process is only invoked if the damage form has not been returned is not representable. Thus, this DFD depicts the data use in processes and by processors but the sequential execution logic expressed by business rules gets lost[49]. Therefore, though the DFD is still a part of many methodologies, it is not considered for a business rule oriented conceptual modeling[50].

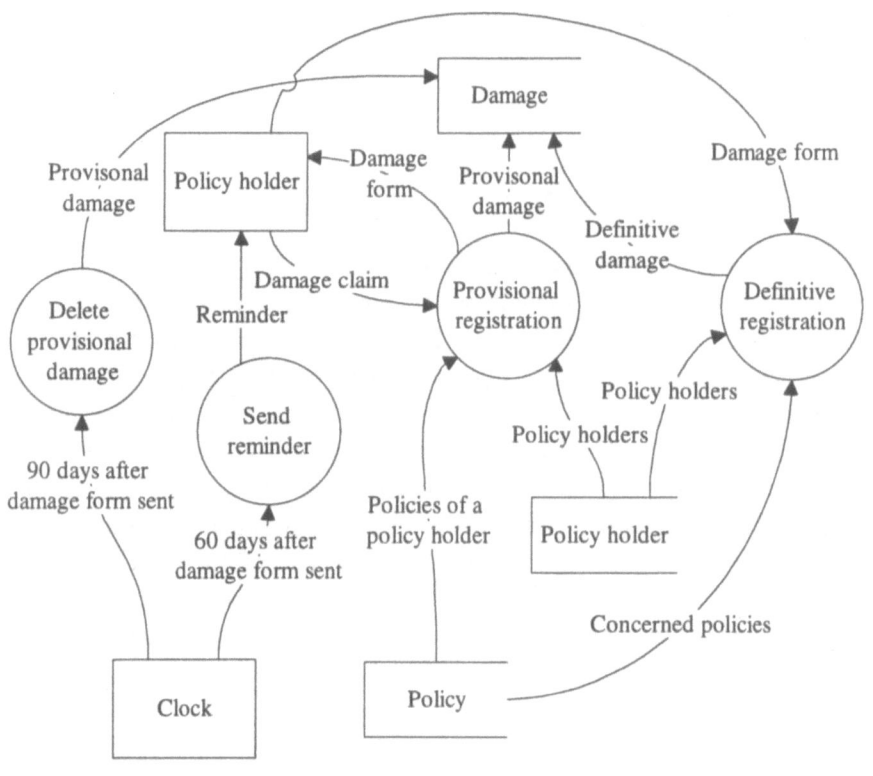

Figure 5-24: DFD for the DOM-2 example

---

[49] This problem could be solved by combining the DFD with e.g. Petri nets, like it is discussed in Franzen/Siegel (1991); however, because in this context only the DFD is regarded, such a combination is not considered.

[50] Further disadvantages of using the DFD are discussed e.g. in Glinz (1991), pp. 25.

5.4.2   Entity-Relationship-Event-Rule Model

5.4.2.1   Introduction

The Entity-Relationship-Event-Rule (ER)^2 model[51] is an enhancement of the En-tity-Relationship-Rule (ER-R) model[52] and was developed at the Computer Sci-ence department of the Georgia Institute of Technology. The ER-R model en-compasses situation-action rules which are combined with the components of an ERM and thus control the states of entities, relationships, and attributes. A situa-tion is defined as a binary tuple of an event and an associated condition; the rule is said to be true or to happen when the event occurs and the condition is satis-fied. The main disadvantage of the ER-R model is the missing representation of the execution logic of rules, i.e., the temporal dependencies between business rules are not representable. Furthermore, with increasing amount of rules the diagram becomes very complex[53]. In the enhancement of the model, the rule part was been replaced by a event-rule net which is similar to the simplified ECA^2 net. Because both components of the (ER)^2 model, i.e., the ERM and the simpli-fied ECA^2 net have already been presented, no further introduction is necessary.

5.4.2.2   Application

For drawing an (ER)^2 model, facts of the submodels *Business rule* and *Data model component* have to be analyzed. The (ER)^2 model can be created either for the entire collection of business rules and the complete ERM or for a specific process only. Considering the amount of business rules in real IS (over 750 in only two transactions of DOM-2), the first alternative will probably fail because of the size of the diagram. Therefore, (ER)^2 models should be drawn for a spe-cific subprocess and thus contain a smaller amount of business rules and only those constructs of the ERM which are referred by these business rules (cf. Figure 5-25).

The (ER)^2 model, as introduced by Tanaka, contains all references between the event-rule net and the accompanying ERM. However, this may lead to a rather complex and confusing diagram. Furthermore, depending on the rule component involved, the relationship between the net and the ERM has different semantics:

- *Event*: User or temporal events cannot refer to modeling constructs. Events raised by data manipulation may result in a directed arc between an event and the data model. This relationship never represents a retrieving or

---

[51]   Tanaka (1992).
[52]   Cf. Tanaka et al. (1991).
[53]   For another application of the ER-R model cf. Herbst et al. (1994), pp. 38.

modifying operation, because an event is a signal which is raised by an attempted or executed action.

- *Condition*: Every condition may encompass references to modeling constructs whose content has to be retrieved for evaluating the condition. The semantics of a directed arc from a condition to the data model thus always represents a retrieval.

- *Action*: Within the two action components, read as well as write operations may lead to a directed arc from the rule symbol to a modeling construct. The semantics of this relationship is therefore a read or a write access.

In order to reduce the complexity of the diagram and not to confuse different semantics, a distinct layer for each component type of a business rules is depicted (cf. Figure 5-25).

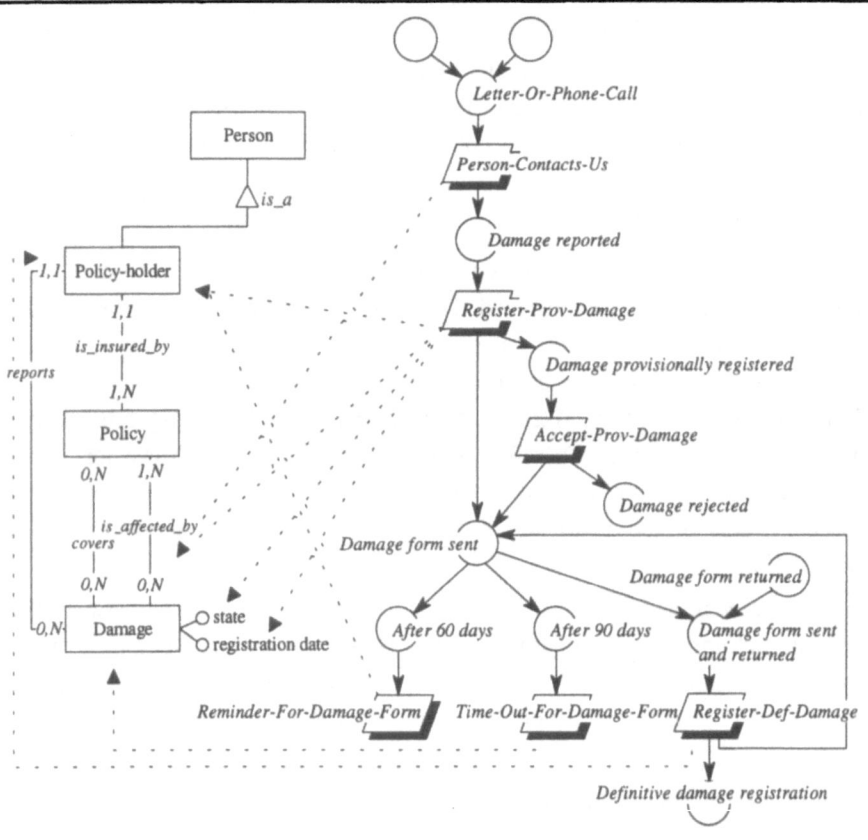

Figure 5-25: (ER)^2 model for the DOM-2 example depicting the „action-layer"

5.5   Data Object Behavior View

In the preceding sections, views on processes, the data model, and on the relationships between both submodels have been regarded. A second dynamic aspect in IS is the behavior of data objects which „concerns life histories of entities. Once entities come into existence, they begin to change. States of entities change in various contexts discretely or continuously until they discontinue their existence. We call this dynamic aspect of entities 'behavior' of 'life history' as a totality of behavior and entities."[54] In contrast to other meta models[55], the *BROCOM* meta model does not represent states of entities by a specific meta entity type. Thus, an entity type which has relevant states is assumed to have an attribute like *State* which may be used for this purpose. As described in Chapter 4, this aspect may be described by using business rules. These have to be formalized according to a specific syntax[56] which allows for an automatic derivation and presentation of entity life cycles. As an example, the life cycle of a *damage* is shown:

Business rule DAMAGE-STATES-1
> *ON*   *(insert damage)*
> *THEN*   *state of damage := 'created'*

Business rule DAMAGE-STATES-2
> *ON*   *(update state of damage)*
> *IF*   *(state of damage ('created') NOT TO ('rejected', 'provisionally accepted', 'accepted'))*
> *THEN*   *reject operation*

Business rule DAMAGE-STATES-3
> *ON*   *(update state of damage)*
> *IF*   *(state of damage ('provisionally-accepted') NOT TO ('rejected', 'accepted'))*
> *THEN*   *reject operation*

Business rule DAMAGE-STATES-4
> *ON*   *(update state of damage)*
> *IF*   *(state of damage('accepted') NOT TO ('closed'))*
> *THEN*   *reject operation*

Business rule DAMAGE-STATES-5
> *ON*   *(update state of damage)*
> *IF*   *(state of damage ('closed') NOT TO ('archived'))*
> *THEN*   *reject operation*

---

[54]   Sakai (1983), p. 111.
[55]   Cf. Section 4.1.4.
[56]   Cf. Section 3.2.2.

Business rule DAMAGE-STATES-6
>   *ON   (delete damage)*
>   *IF   (damage-state NOT IN ('archived', 'rejected'))*
>   *THEN   reject operation*

For the graphical presentation of this view several models exist; some of them have recently been proposed for object-oriented methodologies. In Herbst et al.[57], two representatives were discussed: the entity life history (ELH) diagram as an element of SSADM[58] and the behavior integrate entity relationship (BIER) approach[59]. The ELH combines an entity type with a sequence of allowed states, whereas BIER employs a Petri net for the behavior of entity types. It has been argued that „because ELH is based on a tree structure, whereas BIER uses a net approach, BIER is more flexible to express complex state sequences, e.g. loops"[60]. Thus, the ELH is not further considered. Other well-known techniques are different types of state transition diagrams[61] which are often used in connection with OO approaches[62]; however, because state transition diagrams are regarded as a subtype of Petri nets[63], they are not further considered in this work. Thus, only the combination between ERM and Petri net is discussed in the following.

5.5.1 Petri Net and Entity Relationship Model

5.5.1.1 Introduction

The combination of Petri nets and the ERM has been proposed as ER behavior diagrams[64]. This first approach strongly influenced the development of the BIER approach[65] which has been adapted in order to use it in an object-oriented modeling[66]. However, because in this work an ERM is assumed as a conceptual data

---

[57] Herbst et al. (1994).

[58] Ashworth (1988), p. 157; McDermid (1990); Downs et al. (1992), p. 174; The practical use of ELH's in SSADM is discussed in Edwards et al. (1989), pp. 199. An overview of SSADM is given in Gray/Rao (1993), pp. 85.

[59] Eder et al. (1987); Kappel/Schrefl (1989); Bichler/Schrefl (1994).

[60] Herbst et al. (1994), p. 40.

[61] Davis (1988); Lipeck (1992); Mück (1994).

[62] Cf. Rumbaugh et al. (1991), p. 93; Martin/Odell (1992), pp. 90; Graham (1995), pp. 258; Jacobson et al. (1995), pp. 62. For an overview of methodologies cf. e.g. Hutt (1994); Iivari (1994); Losavio et al. (1994). In Bryant/Evans (1994) OO in general is critically discussed.

[63] Cf. Rosenstengel/Winand (1982), pp. 38.

[64] Cf. Chen (1983), pp. 73; Sakai (1983), pp. 114; Sakai (1990). In Mück (1994) the ERM is combined with statecharts.

[65] Eder et al. (1987); Kappel/Schrefl (1989); Kappel (1991).

[66] Cf. Kappel/Schrefl (1991); Schrefl/Stumpter (1995).

model, the object-oriented data model is not further considered. Furthermore, the ERM as well as Petri nets have been introduced above and are thus not further presented.

5.5.1.2 Application

In Figure 5-26, the life cycle of a *damage* is depicted as specified by the dynamic integrity rules. In the diagram, states are represented by rectangles with the name of the entity type at the top and the state at the bottom[67]. Activities are depicted by transitions. In the BIER approach, the activities are real world actions, e.g., *accept damage*. To derive these real world activities, the formalization of the business rules would have to be adapted: instead of indicating the data manipulation, the accompanying real world activity would be indicated in the event, and rules describing more than one succeeding state would have to be separated. Furthermore, the specification would be 'positive', i.e., not the permissible state changes but the explicit change from one state to the next one are described:

Business rule DAMAGE-STATES-31
> *ON   (accept damage)*
> *THEN   CHANGE state of damage FROM 'provisionally accepted' TO 'accepted'*

Business rule DAMAGE-STATES-32
> *ON   (reject damage)*
> *THEN   CHANGE state of damage FROM 'provisionally accepted' TO 'rejected'*

Because theses modifications of the syntax would lead to major changes in the whole business rule oriented approach, only the 'limited' life cycle model is applied, which includes the data manipulations leading to state changes.

---

[67]   Cf. Schrefl/Stumpter (1995), pp. 5.

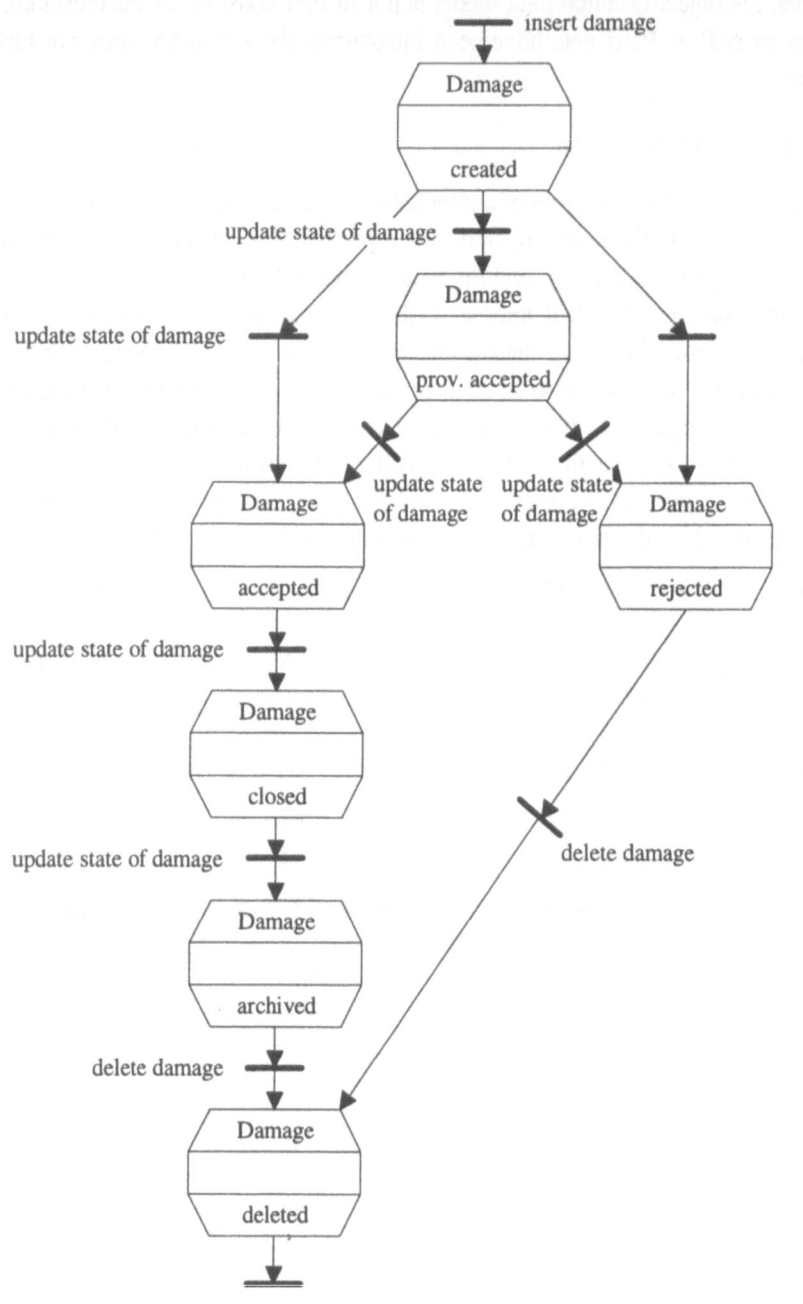

Figure 5-26: Life cycle diagram for entity type damage

5.6   Summary

In Chapter 1 several assumptions for the *BROCOM* approach were made. One of them was that there should be at least one appropriate representation for each view on business rules. Regarding the discussion of models in this chapter, the assumption was correct. Four different views on the *BROCOM* meta model have been discussed and are put together in Figure 5-27.

View on the meta model	Models considered as	
	applicable	not applicable
Process	ECA^2 net (high level Petri net); Simplified ECA^2 net	Conceptual Processing Model
Data	Entity Relationship Model; Nijssen's Integrated Analysis Methodology	
Data use	Entity Relationship Event-Rule model	Data Flow Diagram
Data object behavior	Petri net for entity type/object class	State Transition Diagram; Entity Life History diagram

Figure 5-27: Models for the four views on the business rule meta model

In addition to the views discussed above, analysis of the meta data relevant in the business rule oriented conceptual modeling could result in the graphical presentation of the relationships between

- process and processor types, i.e., the process distribution on manual and automated IS components,
- process and concrete, executing processors, or
- process and organizational units.

Furthermore, the meta data could be analyzed verbally, i.e., by generating reports. As assumed in the introduction, the functionality provided by repository systems ought to be exploited. One major strength of these systems is their effective support for reports based on analyzing the relationships of the meta model in any direction.

6 *BURRO:* A REPOSITORY SYSTEM FOR A BUSINESS RULE-ORIENTED CONCEPTUAL MODELING

In Chapter 4, a meta model and steps of the *BROCOM* approach were introduced. The preceding Chapter 5 encompasses the presentation, application, and evaluation of different models for the graphical presentation of views on the meta model. This chapter describes the implementation of *BURRO,* a business rule repository system which allows the administration of the meta data relevant within the *BROCOM* approach. In this chapter, the implemented *BURRO* system will be applied not only to the example rules used before, but to a large part of the case study DOM-2.

A repository system is in principle a DBMS with some special functionality to administer meta data[1]. This includes the capability of specifying meta models, customizing the user interface, and implementing appropriate verification mechanisms and generators. To cope with the amount of meta data, a repository system has to provide for sophisticated query possibilities which may make use of the classification of the meta data. Furthermore the systems should support semi-graphical and/or graphical representations of query results.

Repository systems have some common functions which cover data definition, data manipulation, query, verification, access control, and version management[2]. Additional functionality result from the modeling approach wherein the repository is to be embedded. Applied to the business rule-oriented conceptual modeling, the employed repository system should be based on the meta model and has to support the modeling steps introduced in the approach (cf. Figure 6-1).

A commercially available product that largely satisfies the requirements and that has been chosen for the implementation of a prototype is the repository system *Rochade[3]*.

---

[1] Cf. Marti (1983), pp. 378; Myrach (1995), pp. 126.
[2] Cf. Myrach (1995), pp. 127.
[3] Cf. R&O (1994a).

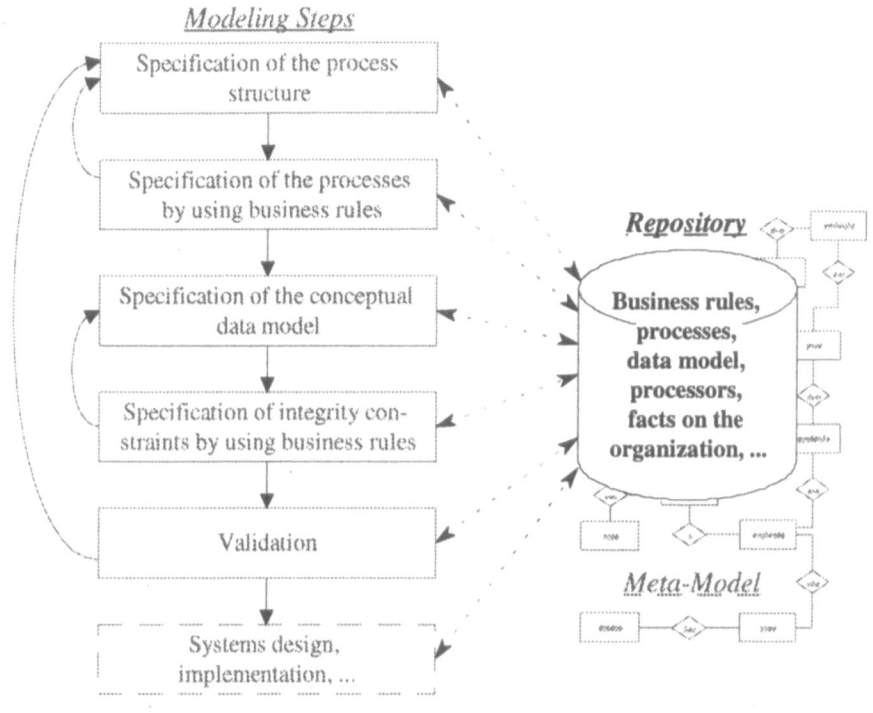

Figure 6-1: Framework of BROCOM: Repository

6.1 Implementation of BURRO

6.1.1 Implementation of the Meta Model

6.1.1.1 Meta Meta Model

The repository system *Rochade* which has been used for the implementation, is itself built upon a data model, the meta meta model (cf. Figure 6-2). This model encompasses the meta meta entity types *Meta model*, *Document type*, *Text attribute*, *Profile*, and *Value attribute*.

Based on this model, individual meta models can be specified and implemented. In *Rochade* these meta models are called information models[4]. An information model consists of a set of meta objects (called document types) which are assigned to a specific instance and are related to each other. The instance of an document type in an information model includes the definition of processing rules (i.e. profiles) and of text and value attributes.

---

[4]   Cf. R&O (1994b), pp. 77.

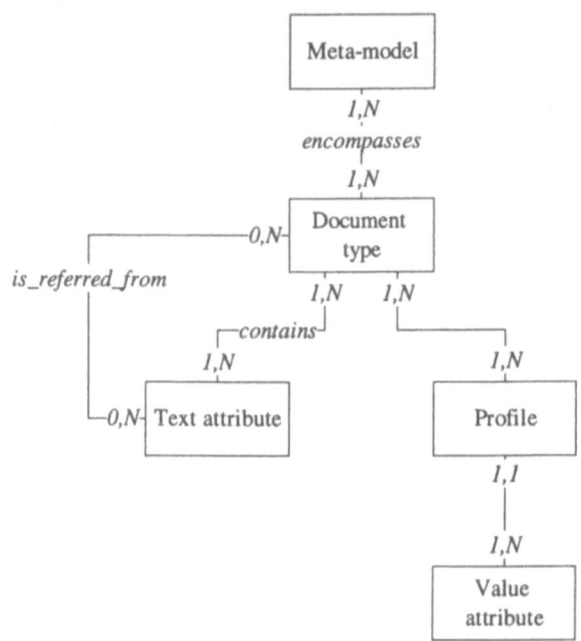

Figure 6-2:   Meta meta model of Rochade[5]

One of the most important parts of the Rochade meta model is the relationship type *is_referred_from* between *Document type* and *Text attribute*. This relationship facilitates the specification of links between any document types which later can be used for analysis of the meta data (e.g. to derive data model components which are referred by business rules).

References in *Rochade* are specified by an arrow, followed by a keyword indicating the referenced document type and the name of an instance of this document type:

→*[Document type] [Instance/document name]*

A reference to a document *Customer* of the the document type *entity-type* would therefore be specified as follows:

→*ENTITY-TYPE Policy-holder*

To accelerate the verbal specification of meta data, a long keyword for the document type can be replaced by a user defined acronym. This may lead to the following reference:

→*ENT Policy-holder*

---

5   Adapted from R&O (1994b), p. 78.

Documents of a specific type which are referred by another document do not have to exist at this moment. *Rochade* distinguishes between active and inactive documents:

- active documents exist as real documents in the repository
- inactive documents are documents that are referred by another document but only exist virtually, i.e., they have not been created explicitly.

Regarding the steps of a business rules-oriented conceptual modeling, these links between document types can be employed as follows: within the specification of business rules for the process description, data model components can be referred from within the rule components using the keywords introduced in Chapter 4. The referred data model components do not have to be dealt with at this moment. For the specification of the data model, all components of this model already exist as *inactive documents* and can be definitely created and modified. If the name of a document is changed, the document names used in the references are changed, too. Thus mistakes in the rule formalization, as the one in the following rule, can be easily corrected:

> ON   ...
> IF   ($\rightarrow$ENT Person is a $\rightarrow$ENT **Polcy-holdr**)
> THEN   ...

In this rule formalization the document name *Policy-holder* is misspelled. During the manipulation of the document *Polcy-holdr* its name is changed to *Policy-holder*. Subsequent the name is changed in *all* references to this formerly misspelled document type

> ON   ...
> IF   ($\rightarrow$ENT Person is a $\rightarrow$ENT **Policy-holder**)
> THEN   ...

6.1.1.2   Physical Data Model

The prototypical implementation of the meta model for business rules focuses on the submodels *Business rule*, *Process* and *Data model components*[6]. Thus, the physical model depicted in Figure 6-3 has been implemented[7].

---

[6]   Cf. the meta model in Chapter 4.
[7]   The detailed description of the implemented document types is given in Appendix B.

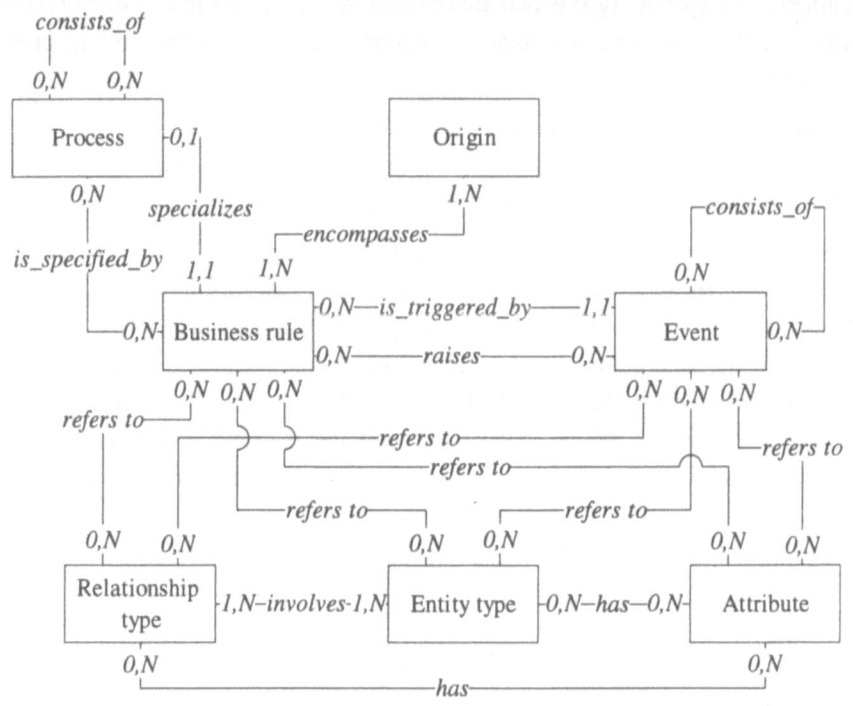

Figure 6-3:   Physical data model for the prototype implementation

6.1.2   Implementation of the User Interface

For the support of the *BROCOM* approach, three main modules have been developed which can be selected in a starting window (cf. Figure 6-4):

- *Processes*: Used for the definition of the process structure (1st step of *BROCOM*)
- *Business rules*: Specification of business rules for describing the processes contained in the process structure (2nd step).
- *Data structure*: Definition of the conceptual data model (3rd step) and the accompanying integrity constraints (4th step).

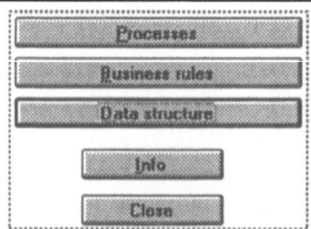

Figure 6-4:   BURRO: Menu selection on the start screen

Processes

The process structure is defined in the window depicted in Figure 6-5. In the middle of the window the current process is displayed; the upper area encompasses its superprocesses and the lower its subprocesses.

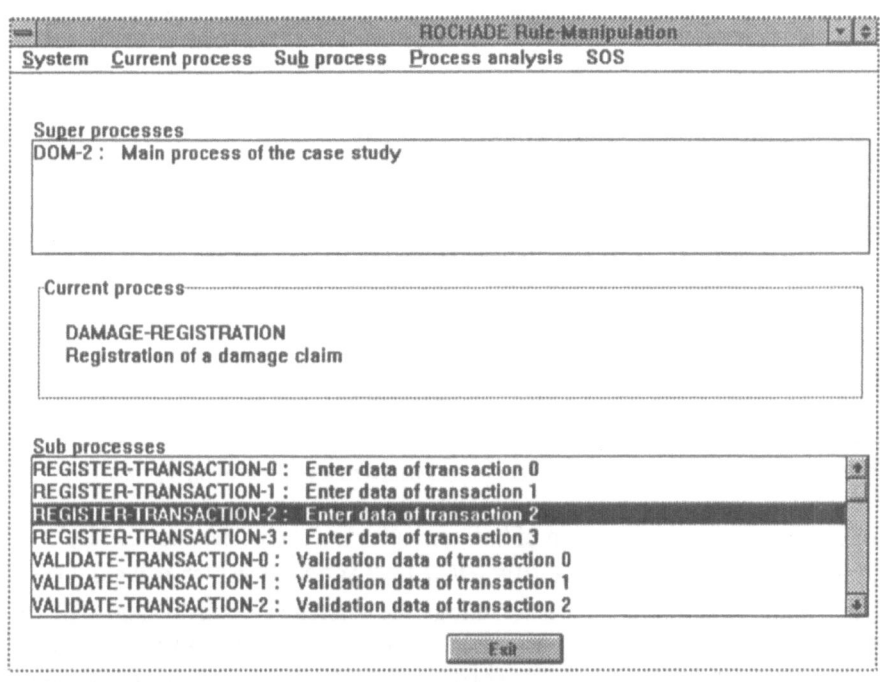

Figure 6-5: *BURRO: Window for defining the process structure*

Business Rules

The administration of business rules specifying processes is done in the window depicted in Figure 6-6. In the figure, those business rules are listed which specify the process *Archive-Damage*.

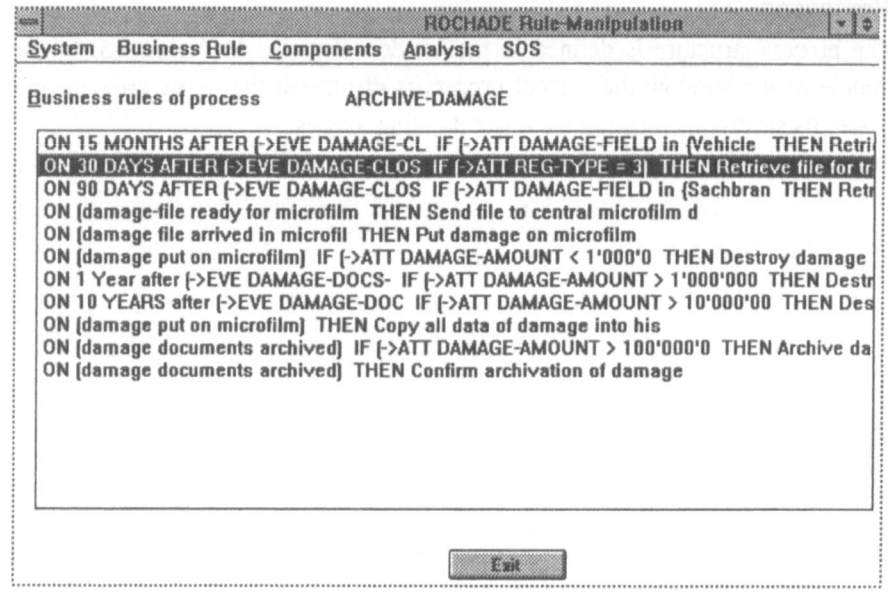

Figure 6-6: **BURRO:** *Window for manipulating business rules describing a process*

The registration or modification of business rules is done in the window depicted in Figure 6-7. The business rule *PERSON-CONTACTS-US,* which is manipulated in the window, shows among other things the references →*ENT DAMAGE* (to the document *DAMAGE* of the type *Entity type*) and →*EVE DAMAGE-REPORTED* (to the document *DAMAGE-REPORTED* of the type *Event,* indicating an event raised in the action).

Figure 6-7: ***Burro:*** *Window for editing a business rule*

Data Structure and Integrity Constraints

Figure 6-8 finally contains four areas for the manipulation of entity types, relationship types, attributes, and business rules. By double-clicking a document displayed in one of the four areas, the user can reduce the documents listed in the other three areas. In Figure 6-8 only those relationships, attributes, and business rules are displayed which make reference to the entity type *DAMAGE.*

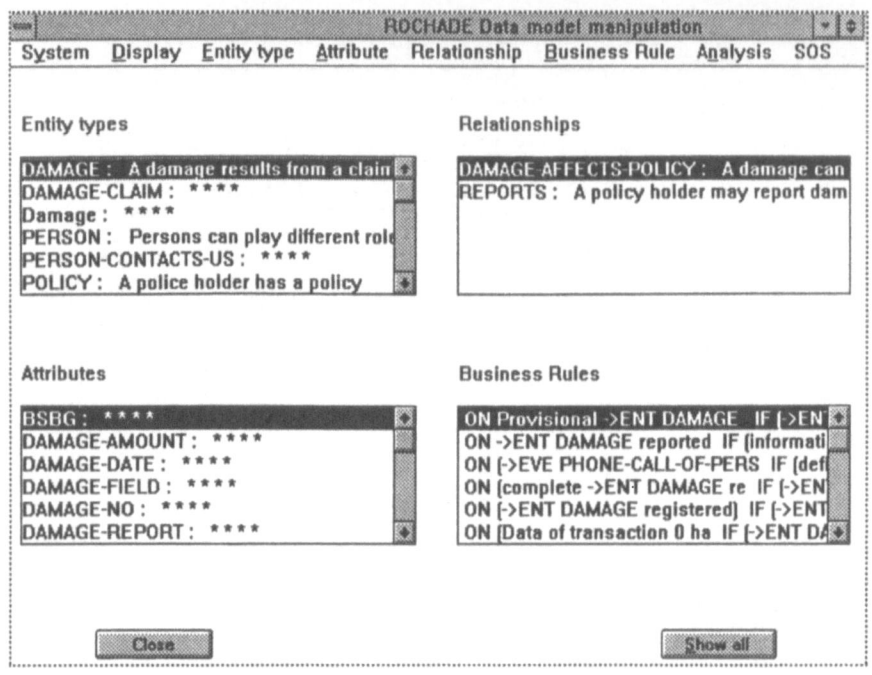

Figure 6-8: **BURRO:** *Window for manipulating the data model and business rules*

6.1.3   Implementation of Graphical Views

In the preceding chapter, four views on the meta model were discussed. Three of them, processes, data structure, and the view on the relationship between them, were selected for the prototypical implementation of the **BURRO** system. The implemented graphical models are[8]

- the process structure diagram,
- the ECA^2 net (as a high-level Petri net) and the simplified ECA^2 net,
- the ERM, and
- the (ER)^2 model.

---

[8]   The resulting diagrams are given in Section 6.2.

6.2 Application of the Repository System to the Case Study „DOM-2"

One goal of the research project was to investigate the applicability of the *BROCOM* approach. Until now, the modeling steps have only been exemplified. They will now be applied to a much larger part of the case study DOM-2. The application concerns the first four steps which are documented with several deliverables in graphical form. All diagrams presented are entirely generated by the repository systems.

6.2.1 Process Structure

The process structure of the case study DOM-2 was already introduced in Chapter 4; thus, only the process structure diagram is depicted (cf. Figure 6-9) as it has been established prior to the specification of each process.

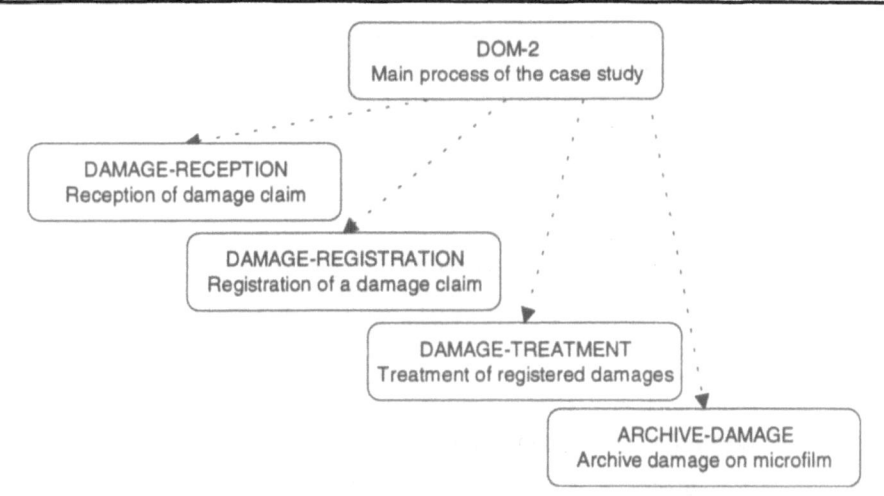

Figure 6-9: Process structure of DOM-2

6.2.2 Process Specification

In order to evaluate the feasibility of the *BROCOM* approach, the main process *DOM-2-Damage-Administration* and its subprocesses are fully specified. Furthermore, the subprocess *Register-Transaction-0* has been specified on a very low level encompassing a large amount of process relevant integrity constraints.

6.2.2.1   Process *DOM-2-Damage-Administration*

The main process *DOM-2-Damage-Administration* divides the damage administration in the four steps damage reception, damage registration, damage treatment, and damage archiving. The process diagrams depicted in Figure 6-10 and
Figure 6-11 are derived from the following business rules:

**** *DAMAGE-RECEPTION* **

    ON   (⇨EVE PHONE-CALL-OF-PERSON) OR
              (⇨EVE LETTER-OF-PERSON)
    IF   (definite ⇨ENT DAMAGE registration necessary)
    THEN   start ⇨ENT DAMAGE registration;
              ⇨EVE DEF-DAMAGE-REGISTRATION
    ELSE   reject damage;
              ⇨EVE DAMAGE-REJECTED

**** *DAMAGE-REGISTRATION* **

    ON   (complete ⇨ENT DAMAGE received)
    IF   (⇨ENT DAMAGE covered by ⇨ENT POLICY)
    THEN   register damage;
              ⇨EVE DAMAGE-REGISTERED
    ELSE   reject damage;
              ⇨EVE DAMAGE-REJECTED

**** *DAMAGE-TREATMENT* **

    ON   (⇨ENT DAMAGE registered)
    IF   (⇨ENT DAMAGE insured by ⇨ENT POLICY)
    THEN   process ⇨ENT DAMAGE claim;
              ⇨EVE DAMAGE-CLOSED
    ELSE   reject damage claim;
              ⇨EVE DAMAGE-REJECTED

**** *ARCHIVE-DAMAGE* **

    ON   (⇨ENT DAMAGE closed)
    THEN   archive ⇨ENT DAMAGE;
              ⇨EVE DAMAGE-ARCHIVED

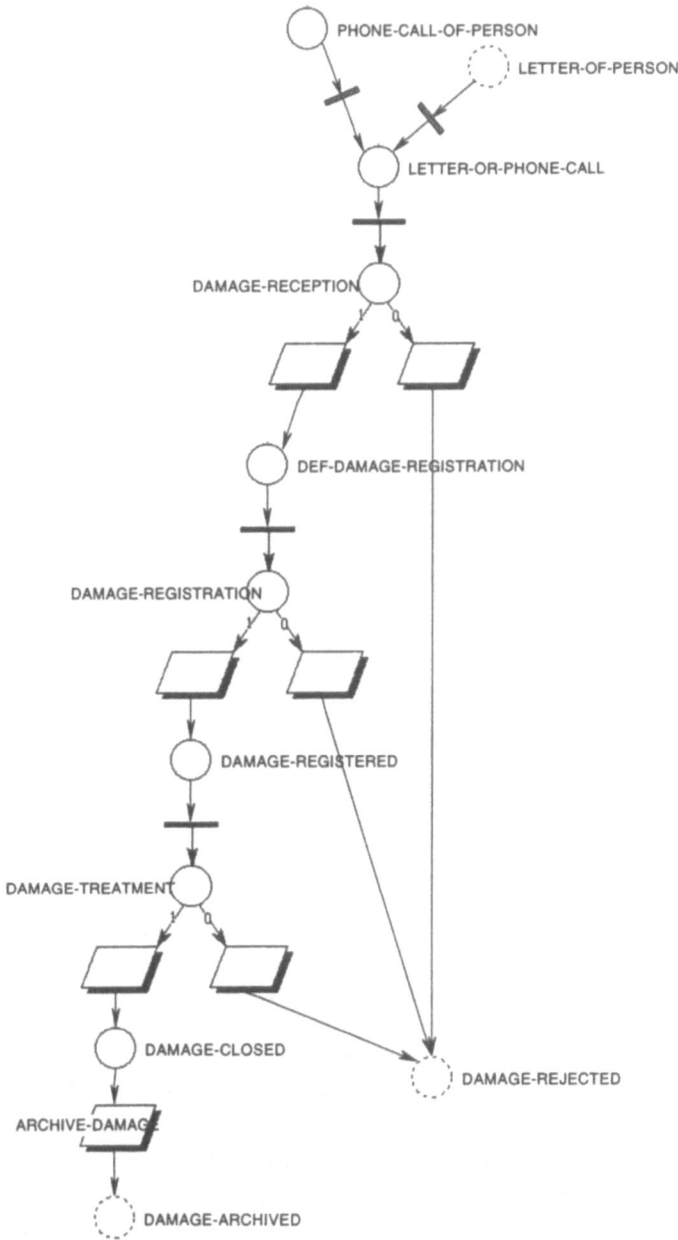

Figure 6-10: ECA² net for the process DOM-2-Damage-Administration

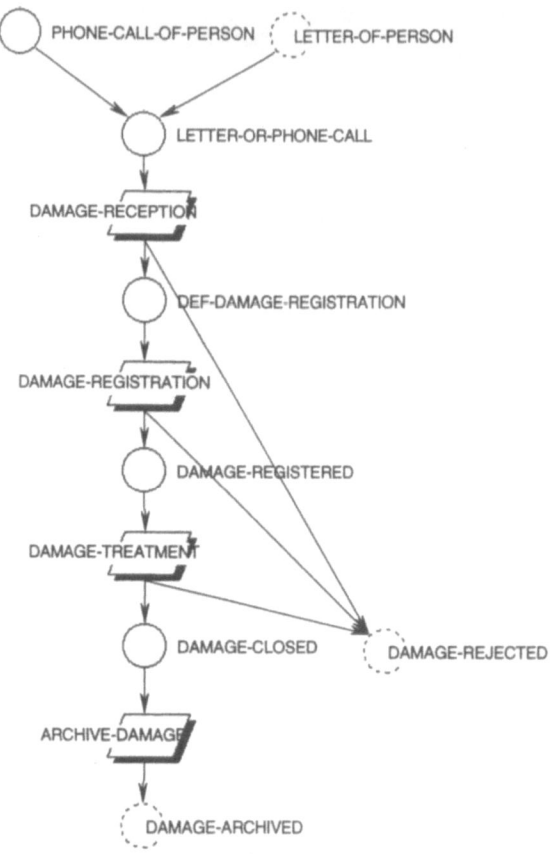

Figure 6-11: Simplified ECA^2 net for the process DOM-2-Damage-Administration

6.2.2.2   Subprocess *Damage-Reception*

The first subprocess of the process *DOM-2-Damage-Administration* is the *Damage-Reception*. It refines the business rule *DAMAGE-RECEPTION* of the main process and therefore starts with the event *LETTER-OR-PHONE-CALL* and leads to the event *DEF-DAMAGE-REGISTRATION*. Between these two events the primarily manual process of receiving a damage claim is described which includes a provisional registration, the sending of a damage form, and time-outs prior to a definitive registration of the damage claim. The process is specified by the following business rules from which Figure 6-12 and Figure 6-13 are derived:

**** *PERSON-CONTACTS-US* **
- ON   (⇨*EVE PHONE-CALL-OF-PERSON*) OR
          (⇨*EVE LETTER-OF-PERSON*)
- IF   (⇨*ENT PERSON is a* ⇨*ENT POLICY-HOLDER*) AND
          (⇨*ENT POLICY-HOLDER wants to report* ⇨*ENT DAMAGE*)
- THEN   begin ⇨*ENT DAMAGE registration;*
          ⇨*EVE DAMAGE-REPORTED*

**** *REGISTER-PROV-DAMAGE* **
- ON   (⇨*ENT DAMAGE reported*)
- IF   (*information about* ⇨*ENT DAMAGE available*)
- THEN   enter available information about ⇨*ENT DAMAGE;*
          ⇨*ATT REGISTRATION-DATE := TODAY;*
          ⇨*EVE PROV-DAMAGE-REGISTERED*
- ELSE   send damage form to ⇨*ENT POLICY-HOLDER;*
          ⇨*EVE DAMAGE-FORM-SENT*

**** *ACCEPT-PROV-DAMAGE* **
- ON   (*provisional* ⇨*ENT DAMAGE registered*)
- IF   (⇨*ENT DAMAGE insured by* ⇨*ENT POLICY*)
- THEN   accept ⇨*ENT DAMAGE provisionally;*
          ⇨*EVE DAMAGE-PROV-ACCEPTED;*
          send damage form;
          ⇨*EVE DAMAGE-FORM-SENT*
- ELSE   reject ⇨*ENT DAMAGE;*
          ⇨*EVE DAMAGE-REJECTED*

**** *REMINDER-FOR-DAMAGE-FORM* **
- ON   (60 days after (⇨*EVE DAMAGE-FORM-SENT*))
- IF   (*damage form not returned*)
- THEN   remind ⇨*ENT POLICY-HOLDER*

**** *TIME-OUT-FOR-DAMAGE-FORM* **
- ON   (90 days after (⇨*EVE DAMAGE-FORM-SENT*))
- IF   (*damage form not returned*)
- THEN   delete provisionally accepted ⇨*ENT DAMAGE*

**** *REGISTER-DEF-DAMAGE* **
- ON   (⇨*EVE DAMAGE-FORM-RECEIVED WITHIN*
          (90 day after (⇨*EVE DAMAGE-FORM-SENT*)))
- IF   (*information in form complete*)
- THEN   ⇨*EVE DEF-DAMAGE-REGISTRATION*
- ELSE   return form to ⇨*ENT POLICY-HOLDER;*
          ⇨*EVE DAMAGE-FORM-SENT*

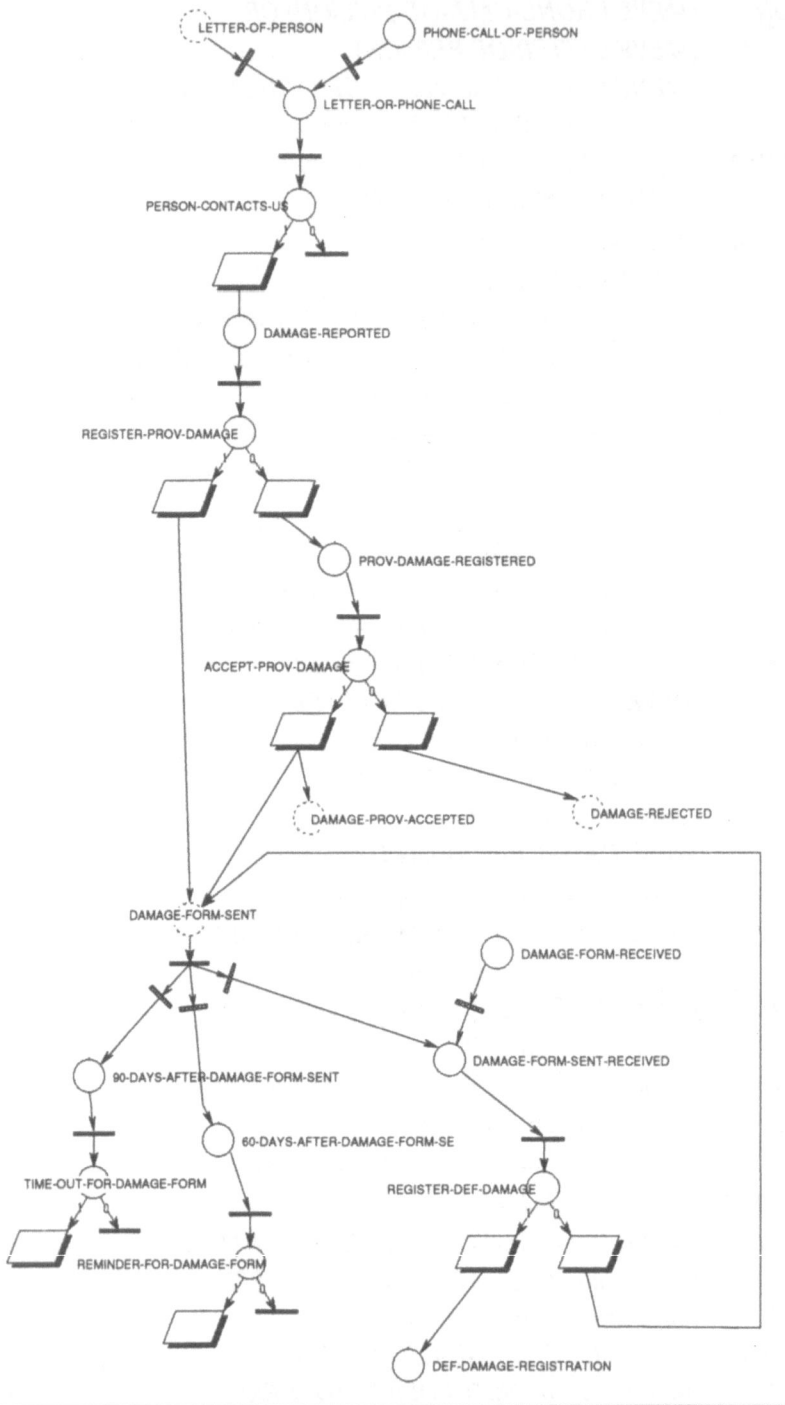

Figure 6-12: ECA² net for the process Damage-Reception

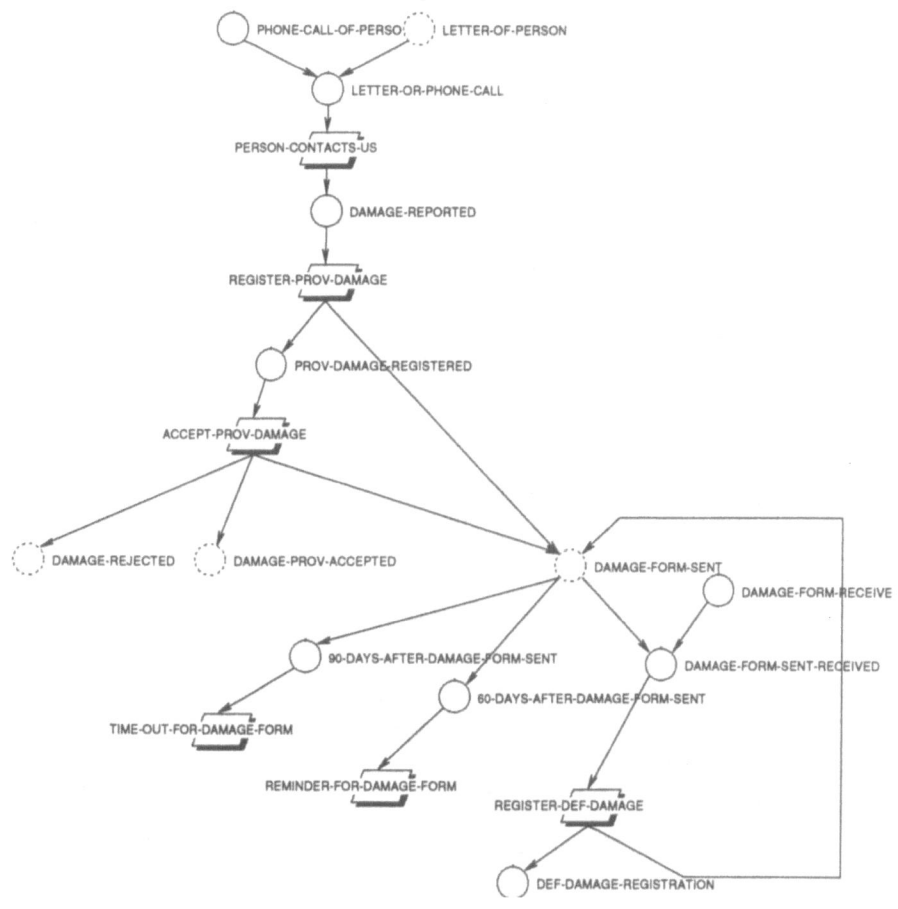

Figure 6-13: Simplified ECA^2 net for the process Damage-Reception

6.2.2.3   Subprocess *Damage-Registration*

The second subprocess of the damage administration encompasses the detailed registration of all information on the damage. The goal of the subprocess is to check whether the damage claim is insured by a policy of the policy holder. The data entry is divided into four transactions which concern four different subjects of data. The rather simple process is specified by the following business rules from which again two process views are derived (cf. Figure 6-14 and Figure 6-16).

**** REGISTER-TRANSACTION-0 **
　　ON　　　(complete ⇨ENT DAMAGE received)
　　THEN　　enter data of transaction 0;
　　　　　　⇨EVE TRANSACTION-0-CONFIRMED

**** VALIDATE-TRANSACTION-0 **
　　ON　　　(data of transaction 0 confirmed)
　　IF　　　(⇨ENT DAMAGE covered by ⇨ENT POLICY)
　　THEN　　validate data of transaction 0;
　　　　　　⇨EVE TRANSACTION-0-VALIDATED
　　ELSE　　reject ⇨ENT DAMAGE;
　　　　　　⇨EVE DAMAGE-REJECTED

**** REGISTER-TRANSACTION-1 **
　　ON　　　(transaction 0 validated)
　　THEN　　enter data of transaction 1;
　　　　　　⇨EVE TRANSACTION-1-CONFIRMED

**** VALIDATE-TRANSACTION-1 **
　　ON　　　(data of transaction 1 confirmed)
　　IF　　　(⇨ENT DAMAGE covered by ⇨ENT POLICY)
　　THEN　　validate data of transaction 1;
　　　　　　⇨EVE TRANSACTION-1-VALIDATED
　　ELSE　　reject ⇨ENT DAMAGE;
　　　　　　⇨EVE DAMAGE-REJECTED

**** REGISTER-TRANSACTION-2 **
　　ON　　　(transaction 1 validated)
　　IF　　　⇨ATT DAMAGE-FIELD in (71..73)
　　THEN　　enter data of transaction 2;
　　　　　　⇨EVE TRANSACTION-2-CONFIRMED
　　ELSE　　no data of transaction 2 needed;
　　　　　　⇨EVE TRANSACTION-2-VALIDATED

**** VALIDATE-TRANSACTION-2 **
　　ON　　　(data of transaction 2 confirmed)
　　IF　　　(⇨ENT DAMAGE covered by ⇨ENT POLICY)
　　THEN　　validate data of transaction 2;
　　　　　　⇨EVE TRANSACTION-2-VALIDATED
　　ELSE　　reject ⇨ENT DAMAGE;
　　　　　　⇨EVE DAMAGE-REJECTED

**** REGISTER-TRANSACTION-3 **
　　ON　　　(transaction 2 validated)
　　THEN　　enter data of transaction 3;
　　　　　　⇨EVE TRANSACTION-3-CONFIRMED

**** *VALIDATE-TRANSACTION-3* **

ON   (data of transaction 3 confirmed)

IF   (⇨ENT DAMAGE covered by ⇨ENT POLICY)

THEN   validate data of transaction 3;

⇨EVE TRANSACTION-3-VALIDATED

ELSE   Reject ⇨ENT DAMAGE;

⇨EVE DAMAGE-REJECTED

**** *END-OF-REGISTRATION* **

ON   (transaction 3 validated)

IF   data complete

THEN   print report;

⇨EVE DAMAGE-REGISTERED

ELSE   complete missing information;

⇨EVE DEF-DAMAGE-REGISTRATION

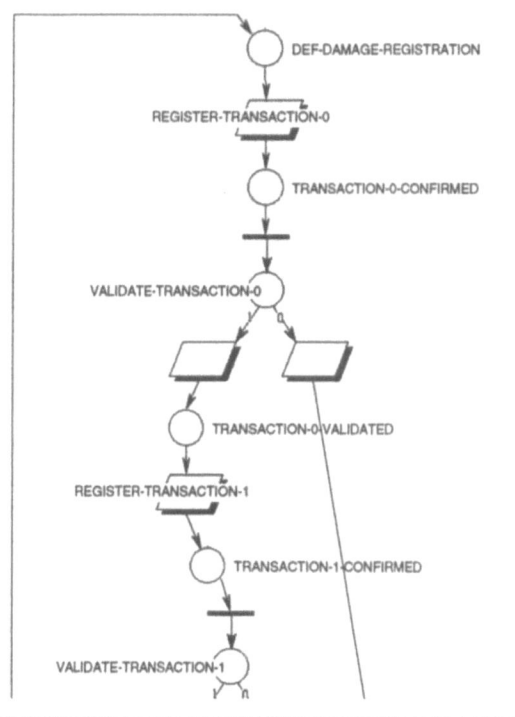

Figure 6-14: ECA² net for the process Damage-Registration *(1ˢᵗ part)*

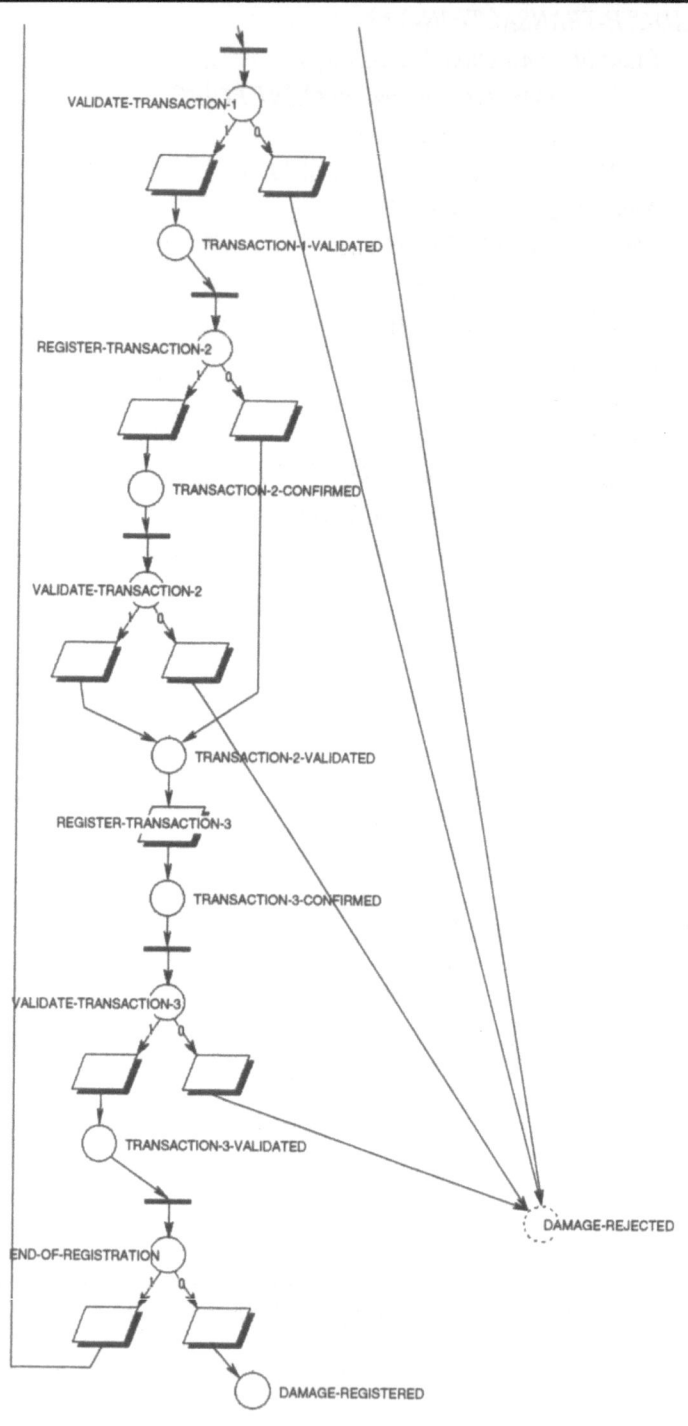

Figure 6-15: ECA² net for the process Damage-Registration *(2nd part)*

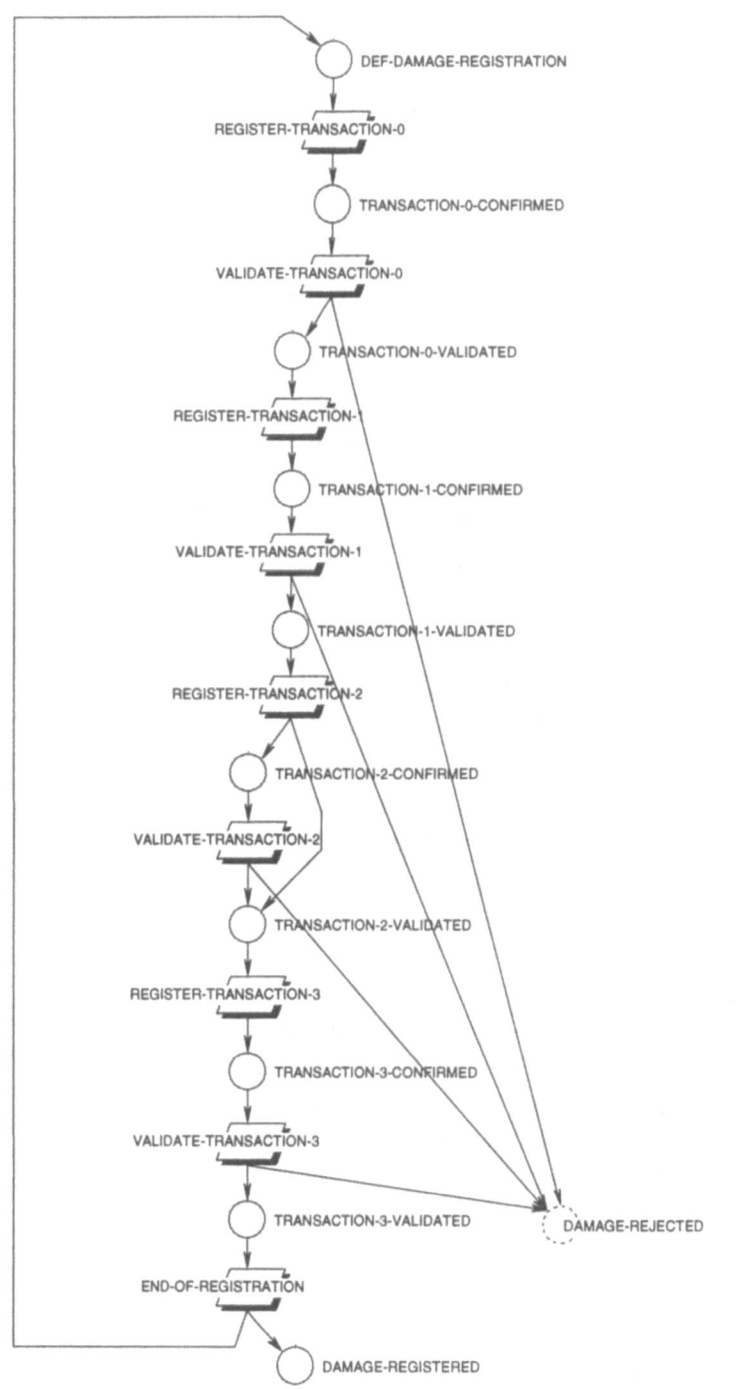

Figure 6-16: Simplified ECA^2 net for the process Damage-Registration

6.2.2.4   Subprocess *Damage-Treatment*

From an organizational point of view, the damage treatment which follows the registration is the most complex subprocess of the damage administration. The first task of the process is the estimation and calculation of the total amount of the damage claim. If appropriate, an advance payment can be invoked at this moment. Because each branch has a specific limit up to which it can handle a damage claim, the responsibility for the damage may be transferred to the central administration if the limit is exceeded. Such a transfer can be either rejected or accepted by the central administration. In the latter case the central administration has to review and perhaps modify the data of the damage. Afterwards the definitive acceptance of the damage claim may take place and is thereby followed by the final payment to the policy holder or other creditors. The process ends with the event *DAMAGE-CLOSED*. Similar to the previous subprocesses, the process is specified using business rules which are used to derive graphic views (cf. Figure 6-17 and Figure 6-18).

**** *DT-MESSAGE* **
    ON   *(⇨ENT DAMAGE registered)*
    IF   *(message to central office necessary)*
    THEN *message on damage to central office*

**** *DT-TREAT-POLICY* **
    ON   *(⇨ENT DAMAGE registered)*
    IF   *(⇨ENT POLICY has to be checked)*
    THEN *invoke policy-check;*
               message to responsible department;
               ⇨EVE POLICY-CHECK

**** *DT-RESPONSIBILITY* **
    ON   *(⇨ATT DAMAGE-AMOUNT calculated)*
    IF   *(⇨ATT DAMAGE-AMOUNT out of branch limit)*
    THEN *transfer damage file to central office;*
               ⇨EVE DAMAGE-TRANSFERRED
    ELSE *treat damage in branch;*
               ⇨EVE DAMAGE-NOT-TRANSFERRED

**** *DT-NO-CLAIM* **
    ON   *(⇨ENT DAMAGE registered)*
    IF   *(⇨ENT POLICY-HOLDER withdraws ⇨ENT DAMAGE-CLAIM)*
               OR
               (⇨ATT DAMAGE-AMOUNT < ⇨ATT POLICY-HOLDER-RISK)
    THEN *close file;*
               ⇨EVENT-TYPE DAMAGE-CLOSED

**** *DT-RE-TRANSFER* **

   ON   (⇨ENT DAMAGE *transferred to central office*)

   IF   (⇨ATT DAMAGE-AMOUNT *out of branch limit*)

   THEN   *accept damage transfer;*

           ⇨EVE DAMAGE-TRANSFER-ACCEPTED

   ELSE   *treat damage in branch;*

           ⇨EVE DAMAGE-NOT-TRANSFERED

**** *DT-CHECK-PAGES* **

   ON   (*damage treatment by central office*)

   THEN   *check and correct pages;*

           ⇨EVE PAGES-DEF-CHECKED

**** *DT-RESERVE* **

   ON   (⇨EVE PAGES-DEF-CHECKED)

   IF   (⇨ATT DAMAGE-AMOUNT > 1000)

   THEN   *enter a reserve*

**** *DT-ACCEPT-DAMAGE* **

   ON   (⇨EVE PAGES-DEF-CHECKED)

   THEN   *accept damage claim;*

           ⇨EVE DAMAGE-CLAIM-ACCEPTED

**** *DT-CALCULATE-DAMAGE* **

   ON   (⇨ENT DAMAGE *registered*)

   THEN   *set* ⇨ATT DAMAGE-STATE := *'Accepted';*

           re-estimate ⇨ATT DAMAGE-AMOUNT;

           ⇨EVE DAMAGE-ESTIMATED

**** *CALCULATE-DAMAGE* **

   ON   (⇨ATT DAMAGE-AMOUNT *estimated*)

   IF   (*all bills and receipts considered*)

   THEN   *finish damage calculation;*

           ⇨EVE DAMAGE-CALCULATED

   ELSE   ⇨ATT DAMAGE-AMOUNT := ⇨ATT DAMAGE-AMOUNT +

           additional amount;

           ⇨EVE DAMAGE-ESTIMATED

**** *ADVANCE-PAYMENT* **

   ON   (⇨ATT DAMAGE-AMOUNT *estimated*)

   IF   (*advance payment appropriate*)

   THEN   *enter advance payment;*

           ⇨EVE ADVANCE-PAYMENT-ENTERED

**** *PAYMENT-RESPONSIBILITY* **

    ON   *(advance payment entered)*

    IF   *(damage treatment by branch) AND*

               (⇨ATT DAMAGE-AMOUNT > ⇨ATT BRANCH-LIMIT)

    THEN *transfer file to central office;*

               ⇨EVE DAMAGE-TRANSFERRED

    ELSE *release advance payment;*

               ⇨EVE ADVANCE-PAYMENT-RELEASED

**** *ADVANCE-PAYMENT-CONTROL* **

    ON   *(advance payment released)*

    IF   *(⇨ATT ENTER-PERS-ID != ⇨ATT RELEASE-PERS-ID)*

    THEN *confirm advance payment;*

               ⇨EVE ADVANCE-PAYMENT-CONFIRMED

    ELSE *no confirmation;*

               ⇨EVE ADVANCE-PAYMENT-ENTERED

**** *FINAL-PAYMENT* **

    ON   *(⇨ENT DAMAGE definitely accepted)*

    IF   *(⇨ATT ADVANCE-PAYMENT < ⇨ATT DAMAGE-AMOUNT)*

    THEN *enter final payment for policy-holder;*

               ⇨EVE FINAL-PAYMENT-ENTERED

    ELSE *close file;*

               ⇨EVE DAMAGE-CLOSED

**** *FIN-PAYMENT-RESPONSIBILITY* **

    ON   *(final-payment entered)*

    IF   *(damage treatment by branch) AND*

               (⇨ATT DAMAGE-AMOUNT > ⇨ATT BRANCH-LIMIT)

    THEN *transfer file to central office;*

               ⇨EVE DAMAGE-TRANSFERRED

    ELSE *release final payment;*

               ⇨EVE FINAL-PAYMENT-RELEASED

**** *CONTROL-FINAL-PAYMENT* **

    ON   *(final payment is released)*

    IF   *(⇨ATT ENTER-PERS-ID != ⇨ATT RELEASE-PERS-ID)*

    THEN *confirm final payment, close file;*

               ⇨EVE DAMAGE-CLOSED

    ELSE *no confirmation;*

               ⇨EVE FINAL-PAYMENT-ENTERED

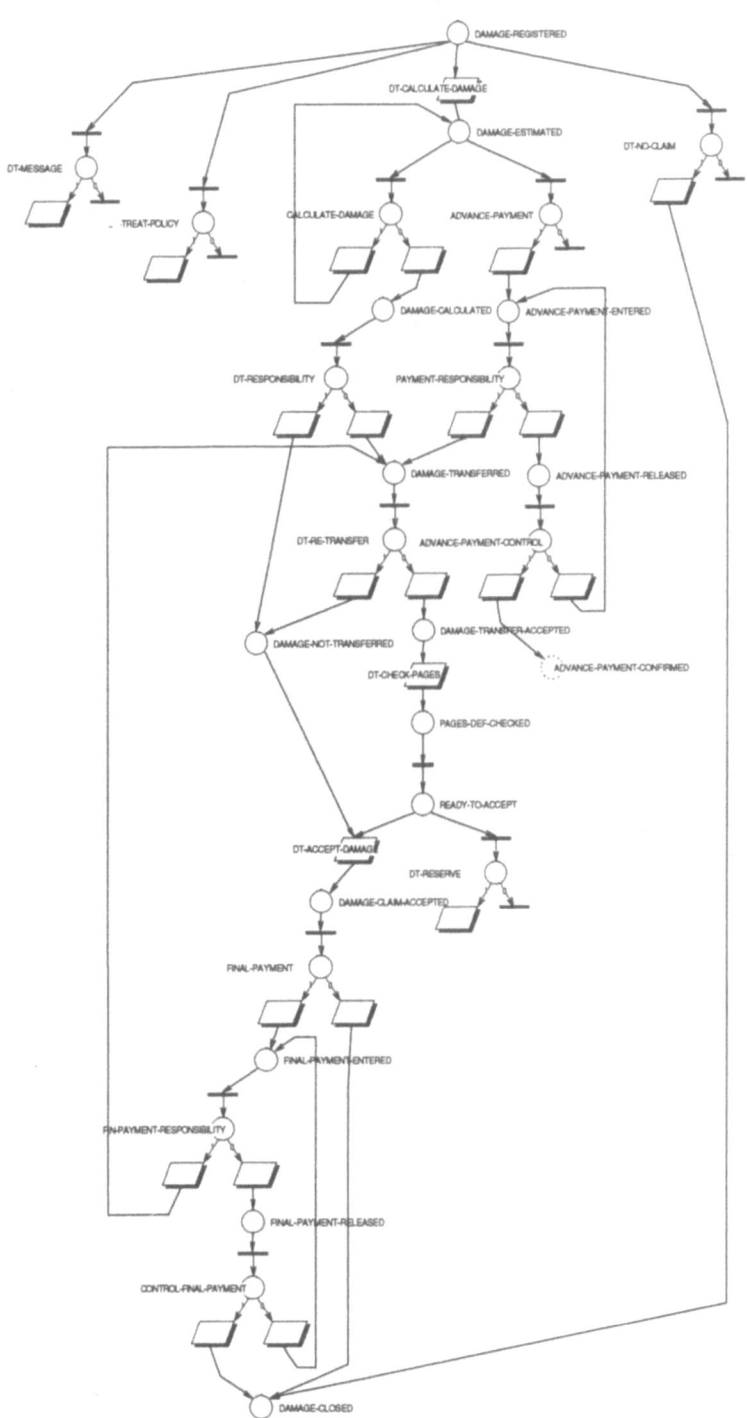

Figure 6-17: ECA^2 net for the process Damage-Treatment

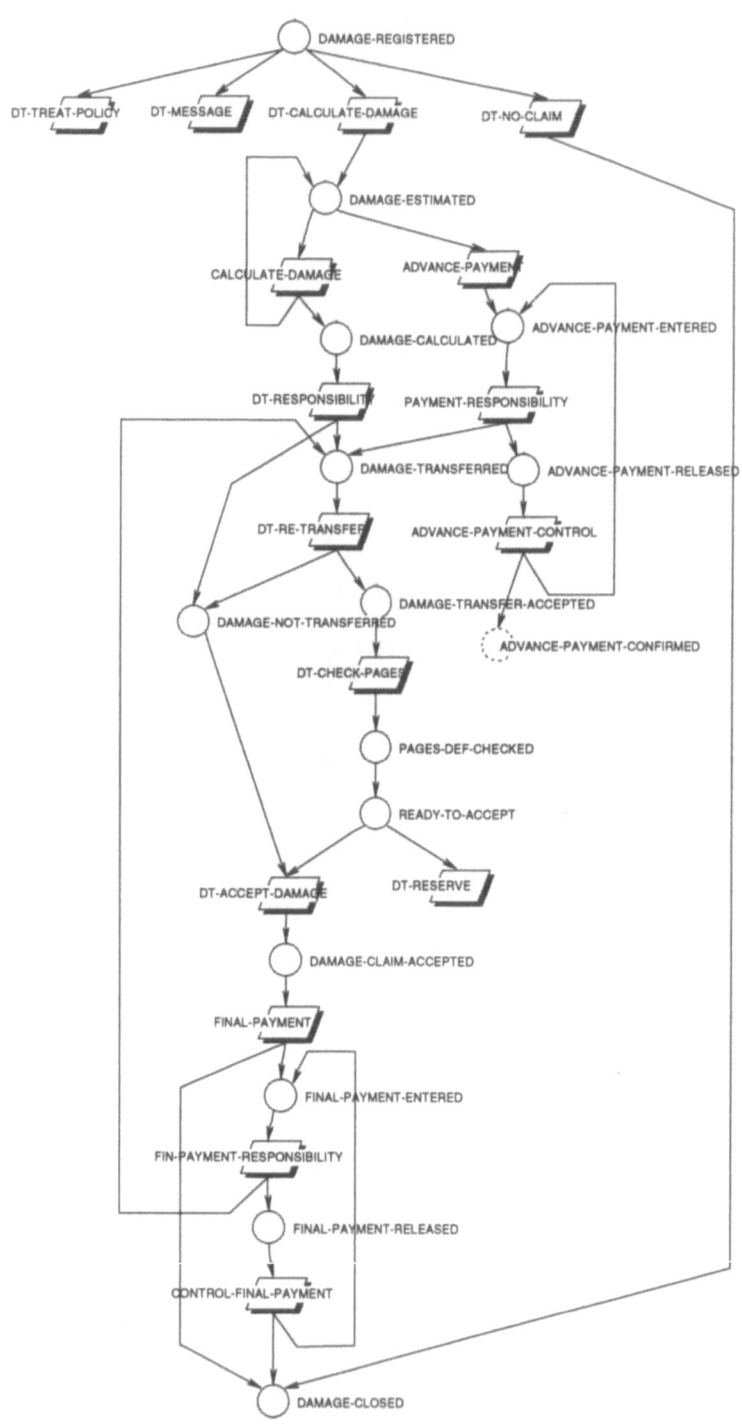

Figure 6-18: Simplified ECA^2 net for the process Damage-Treatment

6.2.2.5   Subprocess *Damage-Archiving*

The subprocess *'Damage-Archiving'* is the refinement of the business rule *'ARCHIVE-DAMAGE'* of the process *'DOM-2-damage-administration'*. The following rules specify the subprocess and serve as a basis for Figure 6-19 and Figure 6-20:

**** *30-DAYS-CLOSED* **

ON	*(30 DAYS AFTER (⇨EVE DAMAGE-CLOSED))*
IF	*(⇨ATT REG-TYPE = 3)*
THEN	*retrieve file for transfer on microfilm;*
	⇨EVE DAMAGE-READY-FOR-MICROFILM

**** *90-DAYS-CLOSED* **

ON	*(90 DAYS AFTER (⇨EVE DAMAGE-CLOSED))*
IF	*(⇨ATT DAMAGE-FIELD = property insurance)*
THEN	*retrieve file for transfer on microfilm;*
	⇨EVE DAMAGE-READY-FOR-MICROFILM

**** *15-MONTH-DAMAGE-CLOSED* **

ON	*15 MONTHS AFTER (⇨EVE DAMAGE-CLOSED)*
IF	*(⇨ATT DAMAGE-FIELD = vehicle insurance)*
THEN	*retrieve file for transfer on microfilm;*
	⇨EVE DAMAGE-READY-FOR-MICROFILM

**** *MICROFILM-DAMAGE* **

ON	*(damage-file ready for microfilm)*
THEN	*send file to central microfilm department;*
	⇨EVE DAMAGE-IN-MICROFILM-DEPT

**** *PUT-DAMAGE-ON-MICROFILM* **

ON	*(damage file arrived in microfilm department)*
THEN	*put damage on microfilm;*
	⇨EVE DAMAGE-ON-MICROFILM

**** *DAMAGE-HISTORY* **

ON	*(damage put on microfilm)*
THEN	*move all data of damage into history database;*
	⇨EVE DAMAGE-DATA-HISTORIZED

**** *ARCHIVE-DOCUMENTS* **
- *ON*   *(damage put on microfilm)*
- *IF*   *(⇨ATT DAMAGE-AMOUNT < 1'000'000.-)*
- *THEN* *destroy damage documents;*
       ⇨EVE ARCHIVED-DOCS-DESTROYED;
       ⇨EVE DAMAGE-ARCHIVED
- *ELSE* *archive damage documents;*
       ⇨EVE DAMAGE-DOCS-ARCHIVED

**** *1-YEAR-ARCHIVED* **
- *ON*   *(1 Year after (⇨EVE DAMAGE-DOCS-ARCHIVED))*
- *IF*   *(⇨ATT DAMAGE-AMOUNT > 1'000'000) AND*
       (⇨ATT DAMAGE-AMOUNT < 10'000'000)
- *THEN* *destroy damage documents;*
       ⇨EVE ARCHIVED-DOCS-DESTROYED

**** *10-YEARS-ARCHIVED* **
- *ON*   *10 YEARS after (⇨EVE DAMAGE-DOCS-ARCHIVED)*
- *IF*   *(⇨ATT DAMAGE-AMOUNT > 10'000'000) AND*
       (⇨ATT DAMAGE-AMOUNT < 100'000'000)
- *THEN* *destroy damage documents*
       ⇨EVE ARCHIVED-DOCS-DESTROYED

**** *DAMAGE-ARCHIVED* **
- *ON*   *(damage documents archived)*
- *THEN* *confirm archiving of damage;*
       ⇨EVE DAMAGE-ARCHIVED

**** *ARCHIVE-FOREVER* **
- *ON*   *(damage documents archived)*
- *IF*   *(⇨ATT DAMAGE-AMOUNT > 100'000'000.-)*
- *THEN* *archive damage documents forever;*
       ⇨EVE DAMAGE-ARCHIVED

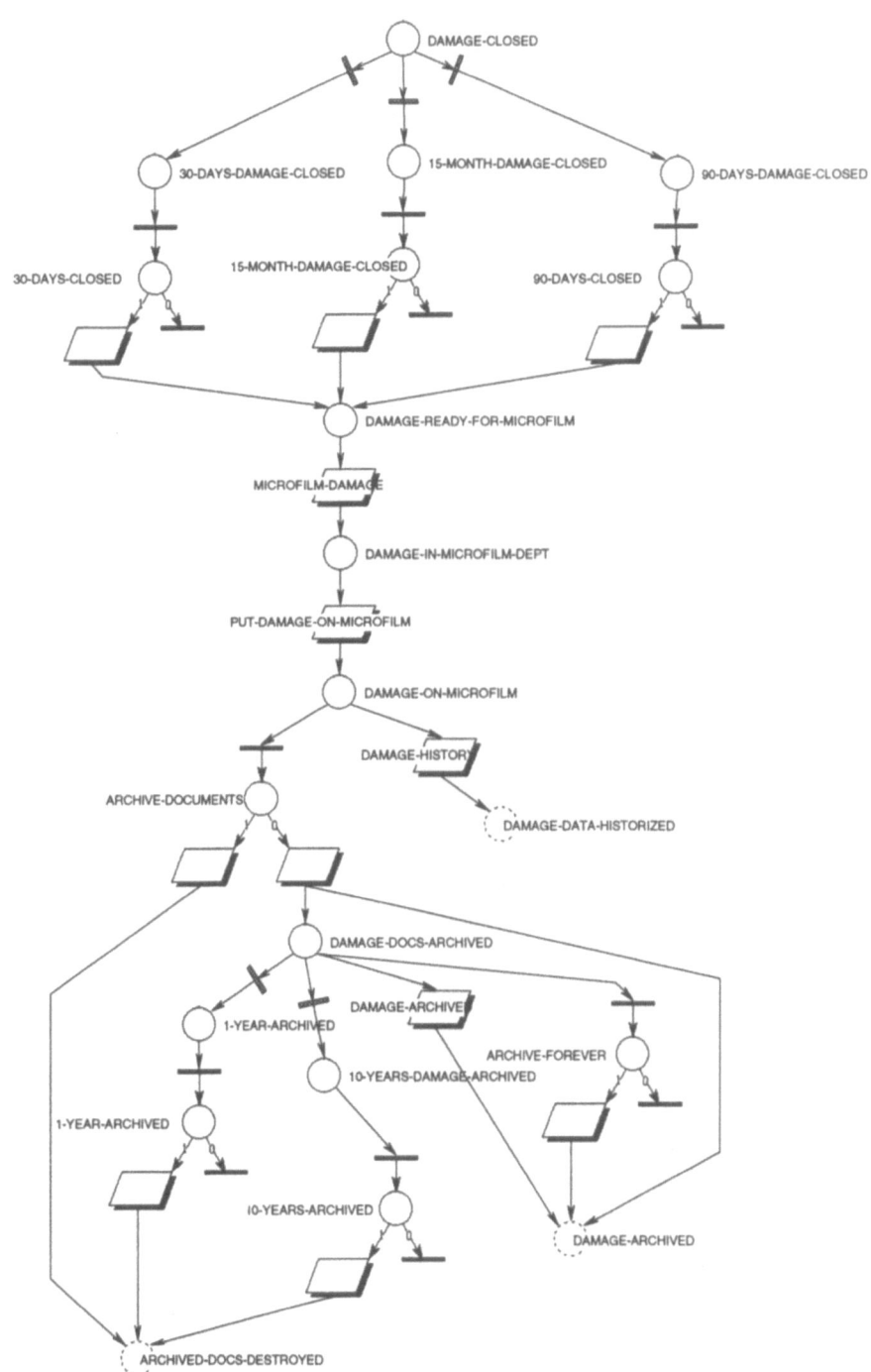

Figure 6-19: ECA^2 net for the process Damage-Archiving

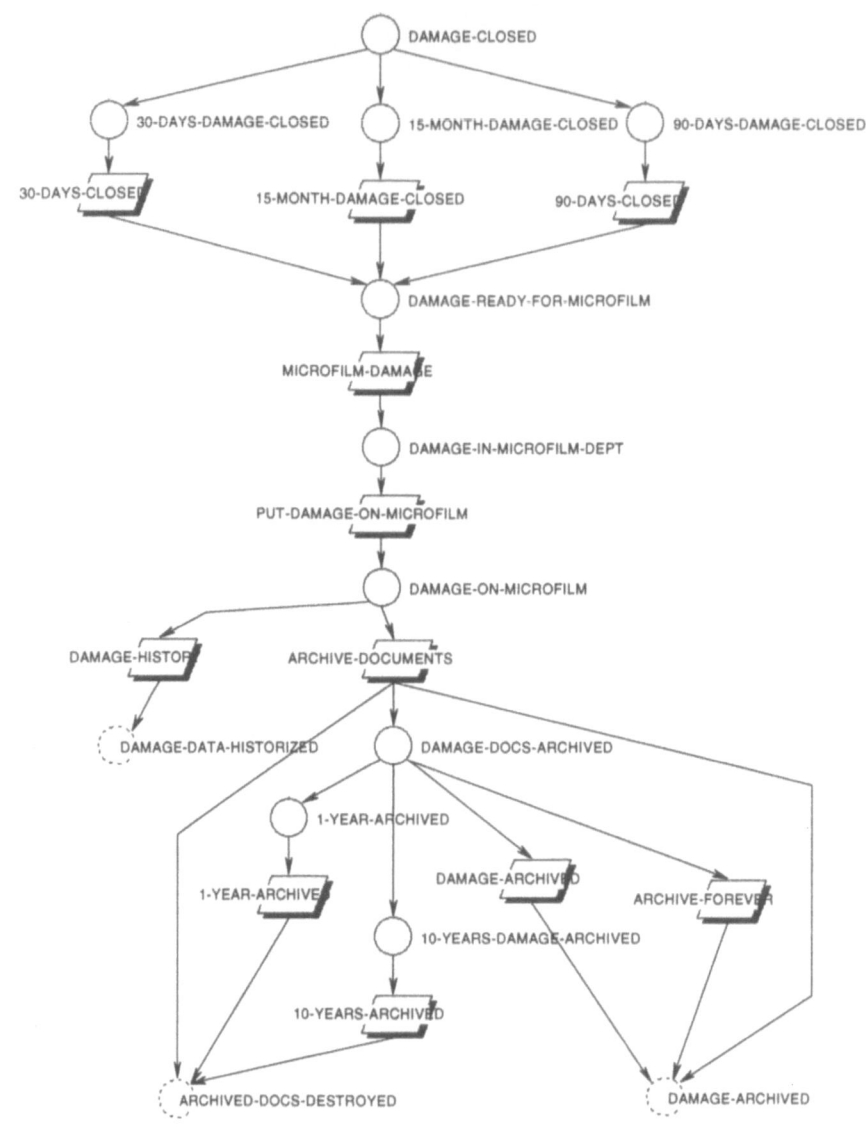

Figure 6-20: Simple ECA^2 net for the process Damage-Archiving

6.2.2.6 Final Process Structure

As already discussed in Chapter 4, the process structure may be extended during the specification of processes. In the application to DOM-2 the final process structure is as depicted in Figure 6-21, whereas the different symbols have the following meaning already mentioned in Chapter 4:

- Process nodes *without shadow* are not specified, i.e., they contain no business rules.
- Process nodes depicted as *rectangles with shadow* are currently being specified.
- Process nodes with *rounded corners and shadow* are completely specified.
- *Dotted lines* are connections which have not yet been assigned to a refined business rule on the higher-level process.
- *'Normal' lines* are connections which are assigned to a specific business rule on the higher-level process.

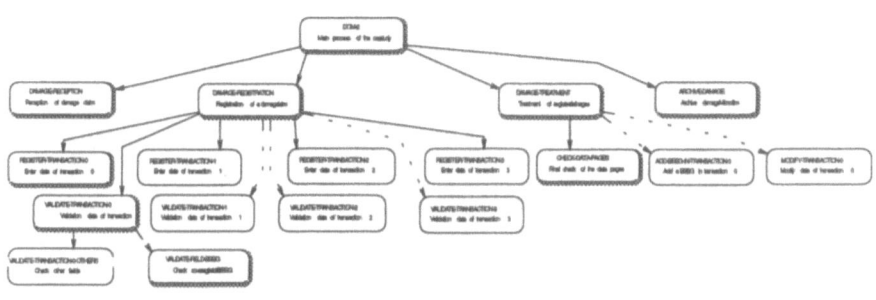

Figure 6-21: Process structure of DOM-2

6.2.3   Data Model

6.2.3.1   Entity-Relationship Model

Prior to the construction of an ERM, all referred components are derived from the business rules listed above. This set of constructs is then structured resulting in the ERM depicted in Figure 6-22.

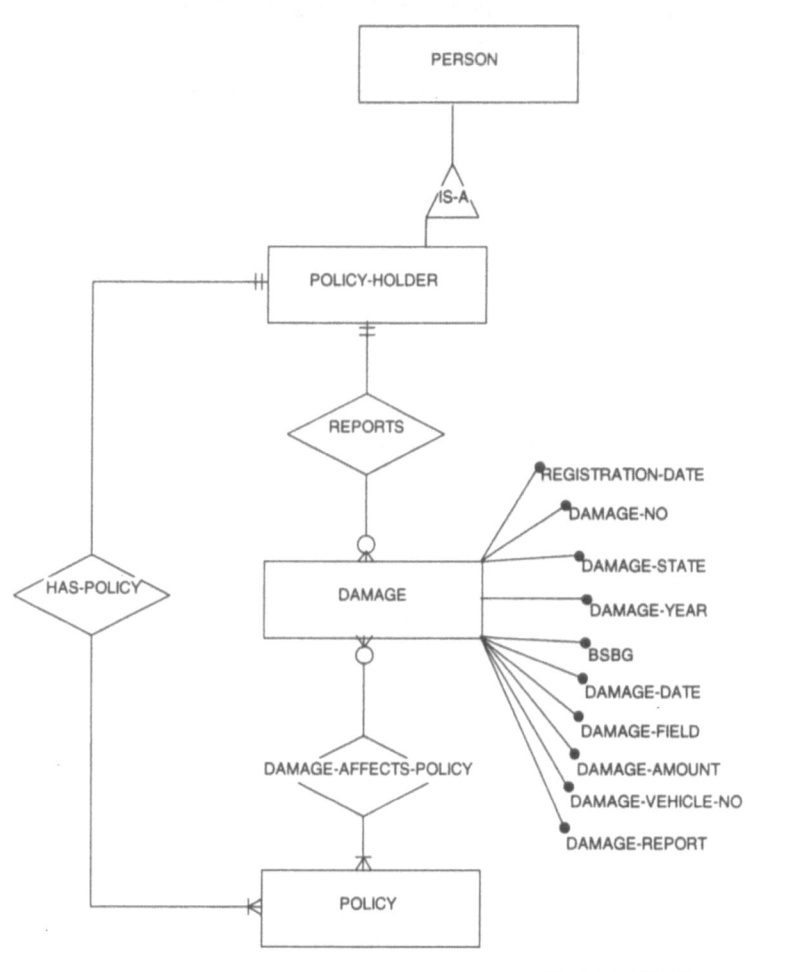

Figure 6-22: ERM for DOM-2

6.2.3.2   Entity-Relationship Event-Rules Model

After having specified the conceptual data model, the interdependencies between business rules and the data model may be analyzed. For this purpose an (ER)^2 model for each process described above may be drawn. However, in the following application only the main process *Damage Administration* is discussed. Figure 6-23 depicts the $(ER)^2$ model for this process including *all* references between the data and the process model. The directed arcs leading from the data model to the parallelograms of the simplified ECA^2 net represent references in the rule condition and those from the rule symbol to the data model symbolize references from within the action. As already discussed in Chapter 5, the representation of all references in a single diagram may result in a confusing diagram.

Figure 6-24 depicts an $(ER)^2$ model for the same process, but only considers references from within the action component, thus leading to a more readable diagram. A further complexity reduction could be achieved by hiding all data model components which are not referred in the selected layer (cf. Figure 6-25).

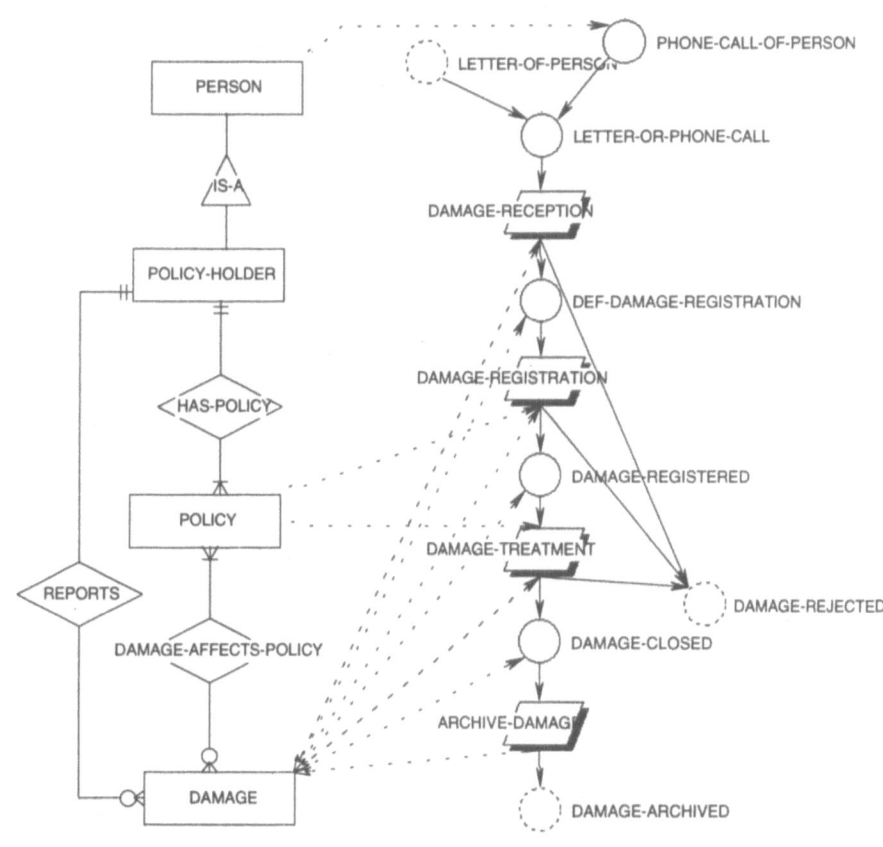

Figure 6-23: $(ER)^2$ model with all references between business rules and the ERM

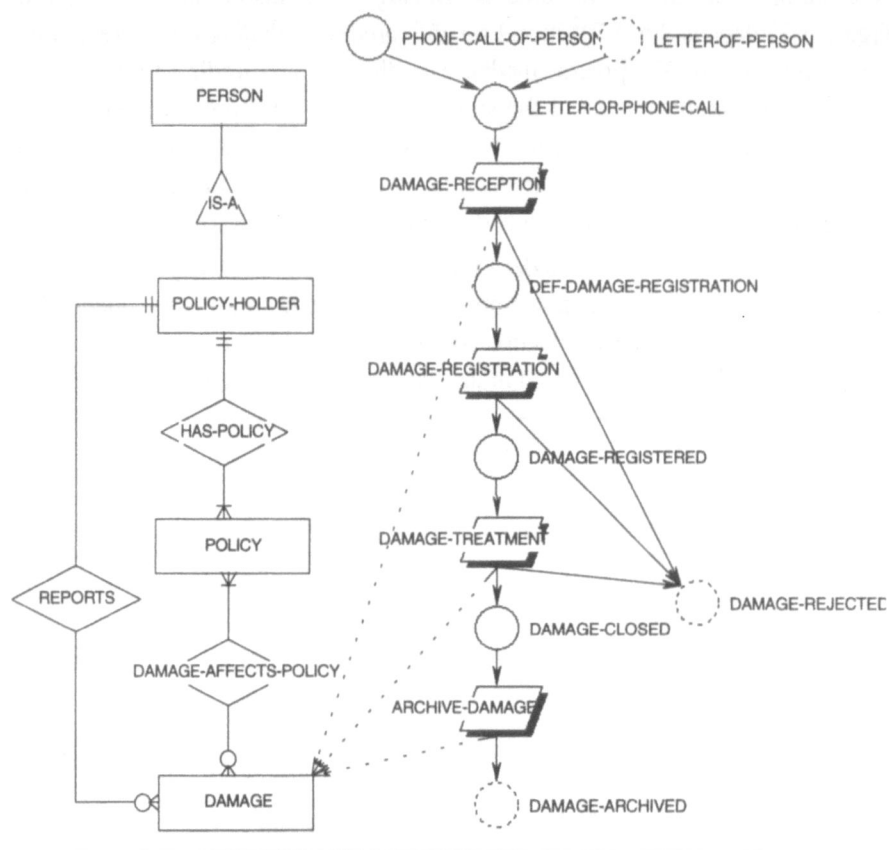

Figure 6-24: (ER)² model with references from rule actions to the ERM

The (ER)² model of Figure 6-24 reveals that within this main process only the
entity type *Damage* gets manipulated, whereas the model in Figure 6-25 shows
that for the evaluation of conditions additionally the entity type *Policy* is needed.

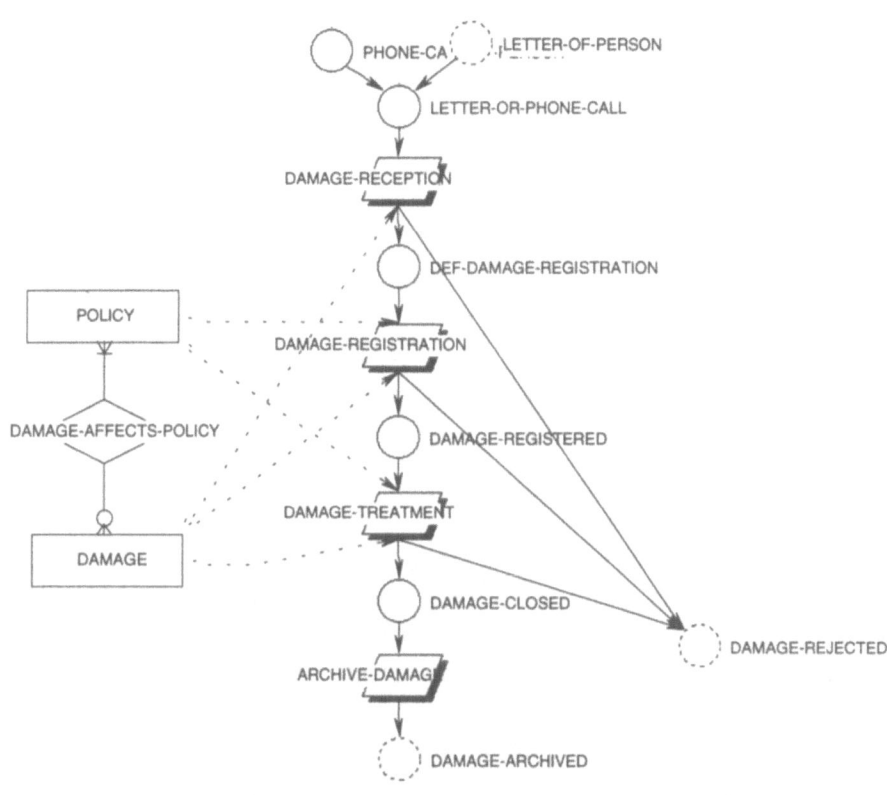

Figure 6-25: (ER)² model with references from rule conditions to the ERM

6.2.4   Integrity Constraints

The formalization of integrity constraints was done during the analysis of the existing IS and resulted in a huge list of approximately 600 business rules. The administration of these rules is supported by the reporting functionality of the repository system. The system can e.g. retrieve all business rules classified as integrity constraints which refer to a specific component of the data model. Using this functionality, e.g. all integrity rules for the attribute *damage-field* could be selected and analyzed. However, because no graphical view for integrity constraints is implemented, integrity constraints are not further discussed within this application of the *BURRO* system.

6.3 Summary

In this chapter the implementation of the repository system *BURRO* which supports the conceptual modeling of business rules has been described. The implemented system provides

- a graphical user interface designed for an optimal support of the steps of the *BROCOM* approach,
- several graphical presentations of views on the *BROCOM* meta model, and
- manifold reporting possibilities.

In order to evaluate the applicability of the *BROCOM* approach, the *BURRO* system was applied to a large part of the case study DOM-2 involving approximately 150 business rules. The main results are as follows:

- All types of business rules found in the case study were adequately formalized in the ECA^2 structure.
- The complexity of the rule set was significantly reduced due to the separate manipulation of different rule types and the refinement of business rules.
- The verbal rule specification in combination with the graphical models presented above, allowed an easy communication with end users in order to validate the rule set. This resulted in finding errors in the processes and finally in correct and valid conceptual specifications.

Thus, this experimental application can be regarded as successful.

7 SUMMARY AND OUTLOOK

The *BROCOM* approach introduced in this work encompasses an orientation towards *business rules* which are not only used for the specification of integrity constraints but also to specify the behavior of processes. The specification of a conceptual model may lead to the *formalization* of business rules; therefore, the aspect of formalization has been discussed from the viewpoint of selected *organizational approaches*. However, though the topic has been discussed since the beginning of industrialization in the 19th century, the appropriate coexistence of non-formalized and formalized rules in an organization is still a controversial topic. This influences the development of an IS because an IS has to guarantee the necessary flexibility for efficient support of an organization's business. Thus, only those facts, which should be formalized from an organizational point of view ought to be fully automated.

The discussion of the organizational theory has been complemented by another topic which is very different but also closely related to business rules. The *implementation alternatives* evaluated in the second part of Chapter 2 considered programming languages of the third, fourth, and fifth generation and rules in deductive and active DBMS. The comparison revealed that ECA rules of active DBMS provide the best functionality for the implementation of business rules. However, the active mechanisms provided by commercially available DBMS still have very restricting shortcomings, e.g., the missing support of composite events and the immature administration functionalities. Thus, programming languages encompassing event constructs also may be used for the implementation of business rules (e.g. certain 4GLs or rule-based 5GLs).

The third chapter focused on business rules in order to provide a basis for the *BROCOM* approach. First, case studies on business rules were presented, of which the case study DOM-2 was presented more closely. In this case study a large insurance IS and the business process in which it is used were analyzed. The result was a collection of over 750 business rules; 130 of these rules describe the business process and 620 prescribe which damages are covered by policies of a policy holder. The second part of this chapter introduced a classification system of integrity constraints and business rule components which is among other things used as a basis for the modeling steps of the *BROCOM* approach. Finally, business rules were considered in their organizational environment, encompassing business processes, origins, processors, and organizational units.

The presentation of the *BROCOM* approach was the subject of the fourth chapter. The introduction was done in two main parts: the *meta model* and the *modeling steps*. The *BROCOM* meta model encompasses business rules structured as ECA^2 as the central submodel. This submodel is complemented by meta entity types for the representation of processes, ERM, processors, origins, and organizational units. The modeling steps in which the meta data of the *BROCOM* approach is manipulated consists of the five steps

1. process structure,
2. process specification,
3. data model,
4. integrity constraints, and
5. validation.

In order to illustrate these steps, they were directly applied to a small example taken from the case study DOM-2. The manipulation of the meta data in the modeling steps is done verbally. As assumed in the introduction, a graphical model ought to be used for the graphical representation of meta data. Chapter 5 therefore dealt with the evaluation of appropriate models for the graphical representation of views on the meta data. The views considered were a process view, a data view, a view on the data use in processes, and a data object behavior view. For each of these views at least one model has been determined which appropriately illustrates the meta data involved.

Another assumption for the development of the *BROCOM* approach was the use of a repository to administer the probably large quantities of meta data. The design and implementation of *BURRO*, a business rule repository system, was presented in the beginning of Chapter 6. The main part of this chapter illustrated the practical application of the *BROCOM* approach by using a large part of the case study DOM-2. This application of *BURRO* included all modeling steps and the representation of all implemented graphical views on the meta data.

The experimental application revealed that the *BROCOM* approach is feasible and the complexity of the rule collection can be sufficiently reduced. The complexity reduction was mainly achieved by the refinement of business rules and the implementation of graphical models. Thus, the current state of *BROCOM* is a prototype which has been successfully applied to a large problem. To gain information on shortcomings of the approach, it should now be applied in practice. Future work may include the enhancement of the *BROCOM* approach as well as of the supporting *BURRO* system:

- *BROCOM.* The modeling approach as introduced in this work facilitates the specification of processes and data models (including integrity constraints).

Thus, enhancements will probably enlarge the scope of the *BROCOM* approach on the conceptual modeling of

- *business processes* as a support for BPR and
- *work flows*[1].

In the last ten years, object-oriented approaches for conceptual modeling have gained much interest and are becoming more and more applied in practice. Therefore, the integration of the *BROCOM* approach with object-oriented approaches could be another interesting enhancement. In these approaches mostly the behavior of objects is described. However, because of the simplicity of the meta model of the *BROCOM* approach, it could be easily used within object-oriented analysis and design. Similar to the derivation of data model components, a keyword ⇨METH for method could be used in the rule syntax thus providing the derivation of object classes, properties, and methods as a basis for an object-oriented model.

- *BURRO*: The repository system is currently a minimal prototype which facilitated the testing of the applicability of the *BROCOM* approach. However, before it is used in practice, the system ought to allow for the derivation of the implementation of business rules, e.g., the generation of rules for active DBMS. Furthermore, algorithms for automatically checking the consistency of the meta data are desirable. Finally, the user interfaces could be optimized and should e.g. enable a graphic-oriented manipulation of the meta data.

A part of this outlook (conceptual work flow modeling and the generation of active rules) is already the subject of further research projects being carried out by the research unit 'Information Engineering' of the Institute of Information Systems at the University of Bern.

---

[1] A project for the conceptual modeling of work flows has been initiated in 1996 and is executed in cooperation with the institute of computer science of the University of Zurich. The main goals of this project are (1) the use of *BROCOM* for the modeling of workflow and (2) to connect *BURRO* with the active DBMS SAMOS.

REFERENCES

ANSI (1989)
>ANSI, American National Standard X3.138-1988: Information Resource Dictionary System (IRDS), New York: American National Standard Institute 1989.

ANSI (1993)
>ANSI, Working Draft X3HS-03-091 of SQL3, New York: American National Standard Institute 1989.

ANSI/X3/SPARC (1975)
>ANSI/X3/SPARC, Study Group on Database Management Systems, Interim Report 75-02-08, in: FDT Report of ACM SIGMOD 7 (1975) 2.

Anwar et al. (1993)
>Anwar, E., Maugis, L., Chakravarthy, S., A New Perspective on Rule Support for Object-Oriented Databases, in: SIGMOD RECORD 22 (1993) 5, pp. 99-108.

Appelrath et al. (1993)
>Appelrath, H.J., Behrends, H., Jasper, H., Ortleb, H., Die Entwicklung aktiver Datenbanken am Beispiel der Krebsforschung, in: W. Stucky, A. Oberweis (Eds.), Proceedings of the GI-Fachtagung Datenbank-Systeme für Büro, Technik und Wissenschaft, Berlin et al.: Springer 1993, pp. 74-83.

Appleton (1984)
>Appleton, D.S., Business Rules: The Missing Link, in: Datamation 30 (1984) 16, pp. 145-150.

Appleton (1986)
>Appleton, D.S., Rule-Based Data Ressource Management, in: Datamation 32 (1986) 5, pp. 86-99.

Appleton (1988)
>Appleton, D.S., Second Generation Applications, in: Database Programming & Design 1 (1988) 2, pp. 48-54.

Appleton (1995)
>Appleton, D.S., Business Reengineering with Business Rules, in: V. Grover, W.J. Kettinger (Eds.), Business Process Change - Reengineering Concepts, Methods and Technologies, Harrisburg/London: Idea Group 1995, pp. 291-329.

Argyris (1957)
>Argyris, Ch., Personality and Organization, New York 1957.

Argyris (1962)
>Argyris, Ch., Interpersonal Competence amd Organizational Effectiveness, Homewood 1992.

Ashworth (1988)
>Ashworth, C.M., Structured Analysis and Design Method (SSADM), in: Information and Software Technology 30 (1988) 3, pp. 153-163.

Atkinson et al. (1989)
Atkinson, M., Bancilhon, A., DeWitt, D., Dittrich, K.R., Maier, D., Zdonik, S., The Object-Oriented Database System Manifesto, in: W. Kim, J.-M. Nicolas, S. Nishio (Eds.), Proceedings of the Conference on Deductive and Object-Oriented Databases, Amsterdam et al.: Elsevier 1989, pp. 223-240.

Azarmi (1992)
Azarmi, N., A Formalisation of Logic Databases and Integrity Constraints, in: U.W. Lipeck, B. Thalheim (Eds.), Modelling Database Dynamics, Berlin et al.: Springer 1992, pp. 133-152.

Balzert (1993)
Balzert, H., Methodenwahl als Basis für den CASE-Erfolg, in: Online (1993) 9, pp. 51-52.

Batini et al. (1992)
Batini, C., Ceri, S., Navathe, S., Conceptual Database Design - An Entity-Relationship Approach, Redwood City et al.: Benjamin/Cummings 1992.

Bauzer-Medeiros/Pfeffer (1991a)
Bauzer-Medeiros, C., Pfeffer, P., A Mechanism for Managing Rule in an Object-oriented Database, Technical Report 65-90, Rocquencourt: GIP Altair 1991.

Bauzer-Medeiros/Pfeffer (1991b)
Bauzer-Medeiros, C., Pfeffer, P., Object Integrity Using Rules, in: P. America (Ed.), ECOOP '91 European Conference on Object-Oriented Programming, Berlin et al.: Springer 1991, pp. 219-230.

Beech et al. (1991)
Beech, D., Bernstein, P., Brodie, M., Carey, M., Gray, J., Lindsay, B., Rowe, L.A., Stonebraker, M., Third-Generation Database System Manifesto, in: R.A. Meersmann, W. Kent, S. Khosla (Eds.), Object-Oriented Databases: Analysis, Design & Construction (DS-4), Amsterdam et al.: Elsevier 1991, pp. 495-511.

Behrends (1995)
Behrends, H., Beschreibung ereignisorientierter Aktivitäten in datanbankgestützten Informationssystemen, PhD Dissertation, Report 3/95, Oldenburg: University of Oldenburg 1995.

Belkhatir/Melo (1994)
Belkhatir, N., Melo, W.L., Tempo: A Support for the Modeling of Objects with Dynamic Behavior, in: A. Verbraeck, H.G. Sol, P.W.G. Bots (Eds.), Proceedings of the 4th International Working Conference on Dynamic Modelling and Information Systems, Delft: Delft University Press 1994, pp. 71-84.

Bell et al. (1990a)
Bell, J., Brooks, D., Goldbloom, E., Sarro, R., Wood, J., Re-Engineering Case Study Recommendations to Applications Developers, Bellevue Golden: US West Information Technologies Group 1990.

Bell et al. (1990b)
Bell, J., Brooks, D., Goldbloom, E., Sarro, R., Wood, J., Re-Engineering Case Study - Analysis of Business Rules and Recommendations for Treatment of Rules in a Relational Database Environment, Bellevue Golden: US West Information Technologies Group 1990.

Bennis (1959)
Bennis, W.G., Leadership Theory and Administrative Behavior, in: Administrative Science Quarterly (1959) 4, p. 259-301.

Bennis (1966)
Bennis, W.G., Organizational Developments and the Fate of Bureaucracy in: Industrial Management Review, 7 (1966), pp. 41-55.

Bennis (1969)
Bennis, W.G., Organization Development, Its Nature, Origins, and Prospects, Reading et al.: Addision-Wesley 1969.

Bernold/Hillenkamp (1988)
Bernold, T., Hillenkamp, U. (Eds.), Proceedings of the International Workshop on Expert Systems in Production and Services, Amsterdam et al.: Elsevier 1988.

Bertino/Apuzzo (1984)
Bertino, E., Apuzzo, D., Integrity Aspects in Data Management Systems, in: Proceedings IEEE Trends and Application Conference Making Databases Work, Los Alamitos: IEEE Computer Society Press 1984, pp. 43-52.

Bertocchi et al. (1993)
Bertocchi, R., De Antonellis, V., Zonta, B., Concepts And Mechanisms For Handling Dynamics in Data Base Applications, in: H.J. Schneider (Ed.), Proceedings of the International Computing Symposium 1983 on Application Systems Development, Stuttgart: B.G. Teubner 1993, pp. 367-381.

Bichler/Schrefl (1994)
Bichler, P., Schrefl, M., Active Object-Oriented Database Design Using Active Object/Behavior Diagrams, in: J. Widom, S. Chakravarthy (Eds.), Proceedings of the 4th International Workshop on Research Issues in Data Engineering, Los Alamitos: IEEE Computer Society Press 1994, pp. 163-171.

Bordoloi/Jenkins (1991)
Bordoloi, B., Jenkins, A.M., The Fourth Generation: An Overview, in: International Journal of Information Resource Management 2 (1991) 1, pp. 24-31.

Branding et al. (1993)
Branding, H., Buchmann, A., Kudrass, T., Zimmermann, J., Rules in an Open System: The REACH Rule System, in: W. Paton (Ed.), Proceedings of the 1st International Workshop on Rules in Database Systems, Berlin et al.: Springer 1993, pp. 111-126.

Brownston et al. (1985)
Brownston, L., Farrell, R. Kant, E. Martin, N., Programming Expert Systems in OPS5, An Introduction to Rule-Based Programming, An Introduction to Rule-Based Programming, Reading et al.: Addison-Wesley 1985.

Bryant/Evans (1994)
Bryant, T., Evans, A., OO oversold, in: Information and Software Technology 36 (1994) 1, pp. 35-42.

Bubenko (1980)
Bubenko, J.A. Jr., Information Modeling in the Context of System Development, in: S.H. Lavington (Ed.), Information Processing 80, Amsterdam et al.: North-Holland 1980, pp. 395-411.

Bubenko/Wangler (1992)
Bubenko, J.A. Jr., Wangler, B, Research Directions in Conceptual Specification Development, in: P. Loucopoulos (Ed.), Conceptual Modeling, Databases, and CASE, New York et al.: John Wiley & Sons 1992, pp. 389-412.

Buchmann/Dayal (1988)
> Buchmann, A.P., Dayal, U., Constraint and exception handling for design, reliability and maintainability, in: R.E. Fulton (Ed.), Proceedings of the ASME Symposium "Engineering Database Management: Emerging issues", New York: ASME 1988, pp. 95-100.

Cacace et al. (1993)
> Cacace, F., Ceri, S., Crespi-Reghizzi, S., Fraternali, P., Paraboschi, S., Tanca, L., The LOGRES prototype, in: SIGMOD RECORD 22 (1993), pp. 550-551.

Carr/Johansson (1995)
> Carr, D.K., Johansson, H.J., Best Practices in Reengineering, New York et al.: McGraw-Hill 1995.

Caseau (1991)
> Caseau, Y., Constraints in an Object-Oriented Deductive Database, in: C. Delobel, M. Kifer, Y. Masunaga (Eds.), Deductive and Object-Oriented Databases, Second International Conference DOOD'91, Berlin et al.: Springer 1991, pp. 292-311.

Ceri (1992)
> Ceri, S., A Declarative Approach to Active Databases, in: F. Golshani (Ed.), Proceedings of the Eighth International Conference on Data Engineering, Los Alamitos: IEEE Computer Society Press 1992, pp. 452-456.

Chakravarthy (1989)
> Chakravarthy, S., Rule Management and Evaluation: An Active DBMS Perspective, in: SIGMOD RECORD 18 (1989) 3, pp. 20-28.

Chakravarthy (1993)
> Chakravarthy, S., A Comparative Evaluation of Active Relational Datebases, Technical Report UF-CIS-TR-93-002, Gainesville: University of Florida 1993.

Chakravarthy (1996)
> Chakravarthy, S., Early Active Database Efforts: A Capsule Summary, in: IEEE Transactions on Knowledge and Data Engineering 7 (1995) 6, pp. 1008-1010.

Chakravarthy et al. (1993)
> Chakravarthy, S., Anwar. E., Maugis, L., Design and Implementation of Active Capability for an Object-Oriented Database, Technical Report UF-CIS-TR-93-001, Gainesville: University of Florida 1993.

Chakravarthy et al. (1994a)
> Chakravarthy, S., Anwar. E., Maugis, L., Mishra, D., Design of Sentinel: an object-oriented DBMS with event-based rules, in: Information and Software Technology 36 (1994) 9, pp. 555-568.

Chakravarthy et al. (1994b)
> Chakravarthy, S., Krishnaprasad, V., Anwar, E., Kim, S.-K., Composite Events for Active Databases: Semantics Contexts and Detection, in: J. B. Bocca, M. Jarke, C. Zaniolo (Eds.), Proceedings of the 20th Conference on Very Large Databases, Los Alamitos: IEEE Computer Society Press 1994, pp. 606-617.

Chakravarthy et al. (1994c)
> Chakravarthy, S., Krishnaprasad, V., Tamizuddin, Z., Badani, R.H., ECA Rule Integration into an OODBMS: Architecture and Implementation, Technical Report UF-CIS-TR-94-023, Gainesville: University of Florida 1994.

Chakravarthy/Mishra (1993)

Chakravarthy, S., Mishra, D., Snoop: An Expressive Event Specification Language For Active Databases, Technical Report UF-CIS-TR-93-007, Gainesville: University of Florida 1993.

Chang/Chang (1982)

Chang, J., Chang, S., Database Alerting Techniques for Office Automation Management, in: IEEE Transactions on Communications 30 (1982) 1, pp. 74-81.

Chen (1975)

Chen, P.P.-S., The Entity-Relationship Model - Toward a Unified View on Data, in: D.S. Kerr (Ed.), Proceedings of the First Conference on Very Large Databases, Los Alamitos: IEEE Computer Society Press 1975, p. 173.

Chen (1976)

Chen, P.P.-S., The Entity-Relationship Model - Toward a Unified View of Data, in: ACM Transactions on Database Systems 1 (1976) 1, pp. 9-36.

Chen (1983)

Chen, P.P.-S., ER - A Historical Perspective And Future Directions, in: C.G. Davis, S. Jajodia, P.-A. Ng, R.T. Yeh (Eds.), Proceedings of the Third International Conference on Entity-Relationship Approach to Software Engineering, Amsterdam et al.: North-Holland 1983, pp. 71-77.

Chen (1992)

Chen, P.P.-S., ER vs. OO, in: G. Pernul, A.M. Tjoa (Eds.), Proceedings of the 11th International Conference on the Entity-Relationship Approach, Berlin et al.: Springer 1992, pp. 1-2.

CODASYL (1973)

CODASYL Data Description Language Committee, CODASYL Data Description Language Journal of Development, in: NBS Handbook 113 (1973).

Codd (1979)

Codd, E.F., Extending the Database Relational Model to Capture More Meaning, in: ACM Transactions on Database Systems 4 (1979) 4, pp. 397-434.

Codd (1990)

Codd, E.F., The Relational Model for Database Management - Version 2, Reading et al.: Addison-Wesley 1990.

Comai et al. (1995)

Comai, S., Fraternali, P., Psaila, G., Tanca, L., A Uniform Model to Express the Behaviour of Rules with Different Semantics, in: M. Berndtsson, J. Hansson (Eds.), Proceedings of the International Workshop on Active and Real-Time Database Systems ARTDB-95, Berlin et al.: Springer 1995, pp. 190-208.

Cooper (1986)

Cooper, P., Expert Systems in Management Science, in: T. Bernold (Ed.), Expert Systems and Knowledge Engineering, Amsterdam et al.: North-Holland 1986, pp. 31-50.

Crozier (1964)

Crozier, M., The Bureaucratic Phenomenon, Chicago: University of Chicago Press 1964.

Date (1981)

Date, C.J., Referential Integrity, in: Proceedings of the Seventh International Conference on Very Large Databases, Los Alamitos: IEEE Computer Society Press 1981, pp. 2-12.

Date (1993)
Date, C.J., A Matter of Integrity, in: Database Programming & Design 6 (1993) 10, pp. 15-18.

Davenport (1993)
Davenport, T.H., Process Innovation - Reengineering Work through Information Technology, Boston: Harvard Business School 1993.

Davis (1988)
Davis, A.M., A Comparison of Techniques for the Specification of External System Behavior, in: Communications of the ACM 31 (1988) 9, pp. 1098-1115.

Davis (1990)
Davis, A.M., Software Requirements Analysis and Specification, Englewood Cliffs: Prentice-Hall 1990.

Dayal (1988)
Dayal, U., Active Database Management Systems, in: C. Beeri, J.W. Schmidt, U. Dayal (Eds.), Proceedings of the 3rd International Conference on Data and Knowledge Bases, San Matheo: Morgan Kaufmann 1988, pp. 150-169.

Dayal (1995)
Dayal, U., Ten Years of Activity in Active Database Systems: What have we learned? in: M. Berndtsson, J. Hansson (Eds.), Proceedings of the International Workshop on Active and Real-Time Database Systems ARTDB-95, Berlin et al.: Springer 1995, pp. 3-22.

Dayal et al. (1988a)
Dayal, U., Buchmann, A.P., McCarthy, D.R., Rules Are Objects Too: A Knowledge Model for an Active, Object-Oriented Database Management System, in: K.R. Dittrich (Ed.), Advances in Object-Oriented Database Systems, Berlin et al.: Springer 1988, pp. 129-143.

Dayal et al. (1988b)
Dayal, U., Blaustein, B., Buchmann, A., Chakravarthy, U., Hsu, M., The HiPAC-Project: Combining Active Databases and Timing Constraints, in: SIGMOD RECORD 17 (1988) 1, pp. 51-70.

De Antonellis/Zonta (1981)
De Antónellis, V., Zonta, B., Modelling Events in Data Base Applications Design, in: Proceedings of the Seventh International Conference on Very Large Databases, Los Alamitos: IEEE Computer Society Press 1981, pp. 23-31.

De Marco (1978)
De Marco, T., Structured Analysis and System Specification, New York: Yourdon 1978.

De Troyer et al. (1988)
De Troyer, O., Meersman, R., Verlinden, P., RIDL* on the CRIS Case: A Workbench for NIAM, in: T.W. Olle, A.A. Verrijn-Stuart and L. Bhabuta (eds.), Computerized Assistance During the Information Systems Life Cycle, Amsterdam et al.: North-Holland 1988, pp. 375-459.

Delcambre/Etheredge (1988)
Delcambre, L.M.L., Etheredge, J.N., The Relational Production Language: A Production Language for Relational Databases, in: L. Kerschberg (Ed.), Proceedings of the Second International Conference on Expert Database Systems, Redwood City et al.: Benjamin/Cummings 1988, pp. 333-350.

Denna et al. (1995)
Denna, E.L., Perry, L.T., Jasperson, J., Reengineering and REAL Business Process Modeling, in: V. Grover, W.J. Kettinger (Eds.), Business Process Change: Concepts, Methods and Technologies, Harrisburg, London: IDEA Group 1995, pp. 350-375.

Denna/Perry (1995)
Denna, E.L., Perry, L.T., Solving Business Problems with Real Integrative Business Solutions, Working Paper, Provo: Brigham Young University's Marriot School of Management 1995.

Diaz et al. (1991)
Diaz, O., Paton, N., Gray, P., Rule Management in Object-Oriented Databases: A Uniform Approach, in: G.M. Lohman, A. Sernadas, R. Camps (Eds.), Proceedings of the Seventeenth International Conference on Very Large Databases, Los Alamitos: IEEE Computer Society Press 1991, pp. 317-326.

Diaz et al. (1994)
Diaz, O., Jaime, A., Paton, N., Al-Qaimari, G., Supporting Dynamic Displays Using Active Rules, in: SIGMOD RECORD 23 (1994) 1, pp. 21-26.

Diaz/Embury (1992)
Diaz, O., Embury, S.M., Generating Active Rules from High-Level Specifications, in: R. Lucas, P.M.D. Gray (Eds.), Proceedings of the Tenth British National Conference on Databases, Berlin et al.: Springer 1992, pp. 228-243.

Dietz (1994)
Dietz, J.L.G., Business Modelling for Business Redesign, in: Proceedings of the Twenty-Seventh Annual Hawaii International Conference on System Sciences, Los Alamitos: IEEE Computer Society Press 1994, pp. 723-732.

Disterer (1988)
Disterer, G., Generationen von Programmiersprachen, Working Paper No. 220, Forschungstelle für Betriebliche Datenverarbeitung und Kommunikationssysteme, University of Kiel 1988.

Dittrich et al. (1986)
Dittrich, K.R., Kotz, A.M., Mülle, J.A., An Event/Trigger mechanism to enforce complex consistency constraints in design databases, in: SIGMOD RECORD 15 (1986) 3, pp. 22-36.

Dittrich (1992)
Dittrich, K.R. (Ed.), Advances in Object-Oriented Database Systems, Berlin et al.: Springer 1988.

Dittrich et al. (1995)
Dittrich, K.R., Gatziu, S., Geppert, A., The Active Database Management System Manifesto, in: T. Sellis (Ed.), Rules in Database Systems, Berlin et al.: Springer 1995, p. 3-35.

Downs et al. (1992)
Downs, E., Clare, P., Coe, I., Structured Systems Analysis And Design Method - Application and Context, 2nd ed., Englewood Cliffs: Prentice-Hall 1992.

Dubois et al. (1986)
Dubois, E., Hagelstein, J., Lahou, E., Ponsaert, F., Rifaut, A., Williams, F., The ERAE Model: A Case Study, in: T.W. Olle, H.G. Sol, A.A. Verijn-Stuart (Eds.), Information Systems Design Methodologies: Improving the Practice, Amsterdam et al: Elsevier 1986, pp. 87-105.

Edelstein (1992)
Edelstein, H., Using Stored Procedures and Triggers, in: DBMS 5 (1992) 9, pp. 66-93.

Eder et al. (1987)
Eder, J., Kappel, G., Tjoa, A.M., Wagner, R.R., BIER - The Behaviour Integrated Entity Relationship Approach, in: S. Spacciapietra (Ed.), Proceedings of the Sixth International Conference on the Entity Relationship Approach, Amsterdam et al.: North-Holland 1987, pp. 147-166.

Eder et al. (1994)
Eder, J., Groiss, H., Nekvasil, H., A Workflow System Based on Active Databases, in: G. Chroust, A. Benczúr (Eds.), Proceedings of the CONectivity '94: Workflow Management - Challenges, Paradigms and Products, Wien/München: Oldenbourg 1994, pp. 249-265.

Edwards et al. (1989)
Edwards, H.M., Thompson, J.B., Smith, P., Results in survey of use of SSADM in commercial and government sectors in UK, in: Information and Software Technology 31 (1989) 1, pp. 21-28.

Eick/Werstein (1993)
Eick, C.F., Werstein, P., Rule-Based Consistency Enforcement for Knowledge-Based Systems, in: IEEE Transactions on Knowledge and Data Engineering 5 (1993) 1, pp. 52-64.

Eisenegger (1994)
Eisenegger, P., Fallstudie zur Analyse von Geschäftsregeln in der Anwendung 'Dezentrale Auftragserfassung Zahlungsverkehr', Masters Thesis, Institue of Information Systems, University of Bern, Bern 1994.

Elmasri/Navathe (1994)
Elmasri, R., Navathe, S.B., Fundamentals of Database Systems, 2nd Ed., Redwood City et al.: Benjamin/Cummings 1994.

Ernst (1988)
Ernst, C.J. (Ed.), Management Expert Systems, Workingham et al.: Addison-Wesley 1988.

ESPRIT Consortium AMICE (1989)
ESPRIT Consortium AMICE (Eds.), Open System Architecture for CIM, Berlin et al.: Springer 1989.

Eswaran (1976)
Eswaran, K.P., Aspects of a Trigger Subsystem in an Integrated Database System, in: Proceedings of the 2nd International Conference on Software Engineering, IEEE Computer Society Press 1976, pp. 243-250.

Eswaran/Chamberlin (1975)
Eswaran, K.P., Chamberlin, D.D., Functional Specifications of a Subsystem for Database Integrity, in: D.S. Kerr (Ed.), Proceedings of the First International Conference on Very Large Databases, Los Alamitos: IEEE Computer Society Press 1975, pp. 48-68.

Etzion (1993)
Etzion, O., PARDES - A Data-Driven Oriented Active Database Model, in: SIGMOD RECORD 22 (1993) 1, pp. 7-14.

Etzioni (1963)
Etzioni, A., Modern organizations, Englewood Cliffs: Prentice-Hall 1963.

Falkenberg (1976)
Falkenberg, E.D., Concepts for Modelling Information, in: G.M. Nijssen (Ed.), Modelling in Database Management Systems, Amsterdam et al.: North-Holland 1976.

Färberböck et al. (1991)
Färberböck, H., Gutzwiller, T., Heym, M., Ein Vergleich von Requirements Engineering Methoden auf Metamodell-Basis, in: M. Timm (Ed.), Requirements Engineering '91 - 'Structured Analysis' und verwandte Ansätze, Berlin et al.: Springer 1991, pp. 40-66.

Ferg (1991)
Ferg, S., Cardinality Concepts in Entity-Relationship Modeling, in: T.J. Teorey (Ed.), Proceedings of the 10th International Conference on the Entity Relationship Approach, San Matheo: Morgan Kaufmann 1991, pp. 1-30.

Ferscha (1994)
Ferscha, A., Qualitative and Quantitative Analysis of Business Workflows using Generalized Stochastic Petri Nets, in: G. Chroust, A. Benczúr (Eds.), Proceedings of the CONectivity '94: Workflow Management - Challenges, Paradigms and Products, Wien/München: Oldenbourg 1994, pp. 222-234.

Fischer (1994)
Fischer, J., Aktive Datenbanken, Working Paper, University of Paderborn 1994.

Franckson (1994)
Franckson, M., The Euromethod Deliverable Model and its contribution to the objectives of Euromethod, in: A.A. Verijn.-Stuart, T.W. Olle (Eds.), Methods and Associated Tools for the Information Systems Life Cycle, Amsterdam et al.: North-Holland 1994, pp. 131-149.

Franzen/Siegel (1991)
Franzen, H., Siegel, G., Die Methode der Strukturierten Analyse mit Petri-Netzen (SA/PN) als Echtzeiterweiterungen, in: M. Timm (Ed.), Requirements Engineering '91 - 'Structured Analysis' und verwandte Ansätze, Berlin et al.: Springer 1991, pp. 178-190.

Freeman/Layzell (1994)
Freeman, M.J., Layzell, P.J., A meta-model of information systems to support reverse engineering, in: Information and Software Technology 36 (1994) 5, pp. 283-294.

Gähler (1987)
Gähler, F., Externe Konsistenzbedingungen: Formulierung und Überwachung, Zürich: ETH Zürich 1987.

Gähler (1991)
Gähler, F., Überwachung von Konsistenzbedingungen, Zürich: Verlag der Fachvereine 1991.

Galbraith (1977)
Galbraith, J.R., Organization Design, Reading et al.: Addison-Wesley 1977.

Gallaire (1988)
Gallaire, H., Bridging the Gap Between AI and Database Logic Approach, in: R.A. Meersmann, A. Sernadas (Eds.), Proceedings of the Second IFIP 2.6 Working Conference on Database Semantics, Amsterdam et al.: North-Holland 1988, pp. 151-172.

Gallaire/Minker (1978)
Gallaire, H., Minkler, J., (Eds.), Proceedings of the Conference on Logic and Databases, New York/London: Plenum Press 1978.

Gallaire et al. (1984)
Gallaire, H., Minkler, J., Nicolas, J.-M., Logic and Databases: A Deductive Approach, in: ACM Computing Surveys 16 (1984) 2, pp. 153-185.

Gane/Sarson (1979)
Gane, C., Sarson, T., Structured Systems Analysis - Tools and Techniques, Englewood Cliffs: Prentice-Hall 1979.

Gatziu (1994)
Gatziu, S., Events in an Active Object-Oriented Database System, Hamburg: Kovac 1994.

Gatziu et al. (1994)
Gatziu, S., Geppert, A., Dittrich, K.R., The SAMOS Active DBMS Prototype, Technical Report 94.16, Institut of Computer Science, University of Zürich 1994.

Gatziu/Dittrich (1993)
Gatziu, S., Dittrich, K.R., Events in an Active Object-Oriented Database System, in: W. Paton and M.H. Williams (Ed.), Proceedings of the First International Workshop on Rules in Database Systems, Berlin et al.: Springer 1993, pp. 23-39.

Gehani et al. (1992)
Gehani, N., Jagadish, H.V., Shueli, O., Event Specification in an Active Object-Oriented Database, in: SIGMOD RECORD 21 (1992) 2, pp. 81-90.

Gehani/Jagadish (1991)
Gehani, N., Jagadish, H.V., Ode as an Active Database: Constraints and Triggers, in: G.M. Lohman, A. Sernadas, R. Camps (Eds.), Proceedings of the Seventeenth International Conference on Very Large Databases, Los Alamitos: IEEE Computer Society Press 1991, pp. 327-336.

Genrich/Lautenbach (1981)
Genrich, H.J., Lautenbach, K., System Modelling with High-Level Petri Nets, in: Theoretical Computer Science 13 (1981), pp. 109-136.

Gerth/Mills (1958)
Gerth, H.H., Mills, C.W., (Eds.), From Max Weber: Essays in Sociology, Oxford: Oxford University Press 1958.

Gertz/Lipeck (1993)
Gertz, M., Lipeck, U.W., Deriving Integrity Maintaining Triggers from Transition Graphs, in: Proceedings of the 9th International Conference on Data Engineering, Los Alamitos: IEEE Computer Society Press 1993, pp. 22-29.

Glasier (1992)
Glasier, P., A Research Project Developing a Technique für Extracting Business Rules from Procedural Code, in: Reverse Engineering Newsletter (1992) 2, p. 6.

Glinz (1991)
Glinz, M., Probleme und Schwachpunkte der Strukturierten Analyse, in: M. Timm (Ed.), Requirements Engineering '91 - 'Structured Analysis' und verwandte Ansätze, Berlin et al.: Springer 1991, pp. 14-39.

Goodall (1985)
Goodall, A., The Guide to Expert Systems, Oxford: Learned 1985.

Graham (1995)
Graham, I., Migrating to Object Technology, Wokingham et al.: Addison-Wesley 1995.

Gray/Rao (1993)
Gray, E.M., Rao, G., Software Requirements Analysis and Specification in Europe: An Overview, in: R.H. Thayer, A.D. McGettrick (Eds.), Software Engineering: A European Perspective, Los Alamitos: IEEE Computer Society Press 1993, pp. 78-96.

Grochla (1978)
Grochla, E., Einführung in die Organisationstheorie, Stuttgart: Poeschel 1978.

Groff/Weinberg (1990)
Groff, J.R., Weinberg, P.N., Using SQL, Berkley et al.: Osborne McGraw-Hill 1990.

Groiss/Eder (1993)
Groiss, H., Eder, J., Einsatz von aktiven Datenbanken für CIM, in: AOV (Ed.), Proceedings of the Vienna IT-Congress "Informations- und Kommunikationstechnologie für das neue Europa", Wien: AOV 1993, pp. 861-870.

Gutenberg (1962)
Gutenberg, E., Unternehmensführung - Organisation und Entscheidungen, Wiesbaden: Gabler 1962.

Gutenberg (1969)
Gutenberg, E., Betriebswirtschaftslehre I, Berlin et al.: Springer 1969.

Haas et al. (1990)
Haas, L.M., Chang, W., Lohman, G.M., McPherson, J., Wilms, P.F., Lapis, G., Lindsay, B., Pirahesh, H., Carey, M.J., Shekita, E., Starburst Mid-Flight: As the Dust Clears, in: Transactions on Knowledge and Data Engineering 2 (1990) 1, pp. 143-160.

Habermann/Leymann (1993)
Habermann, H.-J., Leymann, F., Repository - Eine Einführung, München: Oldenbourg 1993.

Hammer/Champy (1993)
Hammer, M., Champy, J., Reengineering the corporation, New York: Harper Collins 1993.

Hammer/McLeod (1975)
Hammer, M., McLeod, D., Semantic Integrity in a Relational Data Base System, in: D.S. Kerr (Ed.), Proceedings of the First International Conference on Very Large Databases, Los Alamitos: IEEE Computer Society Press 1975, pp. 25-47.

Hammer/McLeod (1976)
Hammer, M., McLeod, D., A Framework for Data Base Semantic Integrity, in: Proceedings of the 2nd International Conference on Software Engineering, IEEE Computer Society Press 1976, pp. 498-504.

Hanks (1992)
Hanks, A., ROSADE: A Methodology for the Extraction of Business Rules, in: Reverse Engineering Newsletter (1992) 2, pp. 6-7.

Hanson (1989)
Hanson, E.N., An Initial Report on the Design of Ariel: A DBMS With an Integrated Production Rule System, in: SIGMOD RECORD 18 (1989) 3, pp. 12-19.

Hanson (1992)

Hanson, E.N., The Design and Implementation of the Ariel Active Database Rule System, Technical Report UF-CIS-TR-018-92, Gainesville: University of Florida 1992.

Hanson/Widom (1993)

Hanson, E.N., Widom, J., An Overview of Production Rules in Database Systems, in: Knowledge Engineering Review 8 (1993) 2, pp. 121-143.

Harrison/Dietrich (1993)

Harrison, J.V., Dietrich, S.W., Integrating Active and Deductive Rules, in: W. Paton (Ed.), Proceedings of the 1st International Workshop on Rules in Database Systems, Berlin et al.: Springer 1993, pp. 289-305.

Hayes-Roth (1988)

Hayes-Roth, F., Knowledge-based Expert Systems: The State of the Art, in: Ernst (1988), p. 3-18.

Heilmann (1994)

Heilmann, H., Workflow Management: Integration von Organisation und Informationsverarbeitung, in: Handbuch der modernen Datenverarbeitung (1994) 176, pp. 8-21.

Herbst (1995)

Herbst, H., A Meta-Model for Specifying Business Rules in Systems Analysis, in: J. Iivari, K. Lyytinen, M. Rossi (Eds.), Proceedings of the Seventh Conference on Advanced Information Systems Engineering (CAiSE '95), Berlin et al.: Springer 1995, pp. 186-199.

Herbst (1996)

Herbst, H., Business Rules in Systems Analysis: A Meta-Model and Repository System, in: Information Systems 21 (1996) 2, pp. 147-166.

Herbst et al. (1994)

Herbst, H., Knolmayer, G., Myrach, T., Schlesinger, M., The Specification of Business Rules: A Comparison of Selected Methodologies, in: A.A. Verijn.-Stuart, T.W. Olle (Eds.), Methods and Associated Tools for the Information Systems Life Cycle, Amsterdam et al.: Elsevier 1994, pp. 29-46.

Herbst/Knolmayer (1994)

Herbst, H., Knolmayer, G., Ansätze zur Klassifikation von Geschäftsregeln, Working Paper 46, Institute of Information Systems, University of Bern 1994.

Herbst/Knolmayer (1995)

Herbst, H., Knolmayer, G., Ansätze zur Klassifikation von Geschäftsregeln, in: Wirtschaftsinformatik 37 (1995) 2, pp. 149-159.

Herbst/Knolmayer (1996)

Herbst, H., Knolmayer, G., Petri Nets as derived process representations in the BROCOM Approach, in: Wirtschaftsinformatik 38 (1996) 4, pp. 391-398.

Herbst/Myrach (1996)

Herbst, H., Myrach, T., A Repository System for Business Rules, to appear in: R. Meersman, L. Mark (Eds.), Proceedings of the Sixth IFIP TC-2 Working Conference on Data Semantics (DS-6), London et al.: Chapman & Hall 1996.

Hess/Brecht (1995)

Hess, T., Brecht, L., State of the Art des Business Process Redesign, Darstellung und Vergleich bestehender Methoden, Wiesbaden: Gabler 1995.

Heym/Österle (1992)

Heym, M., Österle, H., A Reference Model for Information Systems Development, in: K.E. Kendall, K. Lyytinen, J.I. DeGross (Eds), Proceedings of the IFIP WG 8.2 Working Conference on the Impact of Computer Supported Technologies on Information Systems Development, Amsterdam et al.: North-Holland 1992, pp. 215-239.

Hildebrand (1991)

Hildebrand, K., Klassifizierung von Software Tools, in: Wirtschaftsinformatik 33 (1991) 1, pp. 13-16.

Hoffmann et al. (1992)

Hoffmann, W., Scheer, A.W., Backes, R., Konzeption eines Ereignisklassifikationssystems in Prozessketten, Heft 95, Institut für Wirtschaftsinformatik, University of Saarbrücken 1992.

Hoffmann et al. (1993)

Hoffmann, W., Kirsch, J., Scheer, A.W., Modellierung mit Ereignisgesteuerten Prozessketten, Heft 101, Institut für Wirtschaftsinformatik, University of Saarbrücken 1993.

Hofmann (1987)

Hofmann, J., Aktionsorientierte Datenverarbeitung unter besonderer Berücksichtigung des industriellen Fertigungsbereiches, PhD Dissertation, University of Erlangen-Nürnberg 1987.

Horowitz (1992)

Horowitz, B.M., A Run-Time Execution Model for Referential Integrity Maintenance, in: Proceedings of the Eighth International Conference on Data Engineering, Los Alamitos: IEEE Computer Society Press 1992, pp. 548-556.

Hsu et al. (1988)

Hsu, M., Ladin, R., McCarthy, D.R., An Execution Model For Active Data Base Management Systems, in: C. Beeri, J.W. Schmidt, U. Dayal (Eds.), Proceedings of the 3rd International Conference on Data and Knowledge Bases, San Matheo: Morgan Kaufmann Publishers 1988, pp. 171-179.

Hsu/Cheatham (1988)

Hsu, M., Cheatham, T.E. Jr., Rule Execution in CPLEX: A Persistent Objectbase, in: K.R. Dittrich (Ed.), Advances in Object-Oriented Database Systems, Berlin et al.: Springer 1988, pp. 150-155.

Hutt (1994)

Hutt, A.T.F., (Ed.), Object Analysis and Design, New York et al : John Wiley & Sons 1994.

IBM (1980)

IBM (Ed.), Business Systems Planning: Handbuch zur Planung von Informationssystemen, IBM-Form GE 12-1400-1, IBM Germany 1980.

Iivari (1995)

Iivari, J., Object-orientation as structural, functional and behavioural modelling: A comparison of six methods for object-oriented analysis, in: Information and Software Technology 37 (1995) 3, pp. 155-163.

Iivari/Koskela (1993)

Iivari, J., Koskela, E., An Extended EAR Approach For Information System Specification, in: C.G. Davis, S. Jajodia, P.-A. Ng, R.T. Yeh (Eds.), Proceedings of the Third International Conference on Entity-Relationship Approach to Software Engineering, Amsterdam et al.: North-Holland 1983, pp. 605-636.

Inkson et al. (1970)
Inkson, J.H.K., Pugh, D.S. Hickson, D.J., Organization, Context and Structure: An Abbreviated Replication, in: Administrative Science Quarterly (1970), pp. 318-329.

ISO (1989)
ISO, Draft International Standard ISO/IEC DIS 10027, Information Processing Systems - Information Resource Dictionary System (IRDS) Framework, Geneva: International Organization for Standardization 1989.

Jacob/Froscher (1990)
Jacob, R.J.K., Froscher, J.N., A Software Engineering Methodology for Rule-Based Systems, in: IEEE Transactions on Knowledge and Data Engineering 2 (1990) 2, pp. 173-189.

Jacobson et al. (1995)
Jacobson, I., Ericsson, M., Jacobson, A., The Object Advantage - Business Process Reengineering with Object Technology, Wokingham et al.: Addison-Wesley 1995.

Jaeger/Freytag (1995)
Jaeger, U., Freytag, J.C., An Annotated Bibliography on Active Databases, in SIGMOD RECORD 24 (1995) 1, pp. 58-69.

Jakubczik/Skubch (1994)
Jakubczik, G.-D., Skubch, N., Business Process Reengineering: Massstab für den langfristigen IV-Einsatz, in: Online (1994) 4, pp. 58-64.

Jansen/Compton (1989)
Jansen, B., Compton, P., Data dictionary approach to the maintenance of expert systems: The Knowledge Dictionary, in: Knowledge-Based Systems 2 (1989) 1, pp. 15-26.

Jarke/Pohl (1992)
Jarke, M., Pohl, K., Information Systems Quality and Quality Information Systems, in: K.E. Kendall, K. Lyytinen and J.I. DeGross (eds.), The Impact of Computer Supported Technologies on Information Systems Development, Amsterdam et al.: North-Holland 1992, pp. 345-375.

Jones/Walsham (1992)
Jones, M., Walsham, G., The Limits of the Knowable: Organizational and Design Knowledge in Systems Development, in: K.E. Kendall, K. Lyytinen, J.I. DeGross (Eds) Proceedings of the IFIP WG 8.2 Working Conference on the Impact of Computer Supported Technologies on Information Systems Development, Amsterdam et al.: North-Holland 1992, pp. 195-213.

Joosten (1994)
Joosten, S., Trigger Modelling for Workflow Analysis, in: G. Chroust, A. Benczúr (Eds.), Proceedings of the CONectivity '94: Workflow Management - Challenges, Paradigms and Products, Wien/München: Oldenbourg 1994, pp. 236-247.

Jorysz/Vernadat (1990a) Jorysz, H.R., Vernadat, F.B., CIM-OSA Part 1: total enterprise modelling and function view; in: International Journal of Computer Integrated Manufacturing 3(1990)3/4, pp. 144-156.

Jorysz/Vernadat (1990b)
Jorysz, H.R., Vernadat, F.B., CIM-OSA Part 2: information view; in: International Journal of Computer Integrated Manufacturing 3(1990)3/4, pp. 157-167.

Kappel (1991)

Kappel, G., Reorganizing Object Behavior by Behavior Composition - Coping with Evolving Requirements in Office Systems, in: H.-J. Appelrath (Ed.), Proceedings of the GI-Fachtagung Datenbanksysteme in Büro, Technik und Wissenschaft, Berlin et al.: Springer 1991, pp. 446-453.

Kappel et al. (1994a)

Kappel, G., Rausch-Schott, S., Retschitzegger, W., Vieweg, S., TriGS - Making a Passive Object-Oriented Database System Active, in: Journal of Object-Oriented Programming 7 (1994) 4, pp. 40-51.

Kappel et al. (1994b)

Kappel, G., Rausch-Schott, S., Retschitzegger, Beyond Coupling Modes - Implementing Active Concepts on Top of a Commercial ooDBMS, International Symposium on Object-Oriented Methodologies and Systems, Berlin et al.: Springer 1994, pp. 189-204.

Kappel et al. (1994c)

Kappel, G., Rausch-Schott, S., Retschitzegger, W., TriGS$_{flow}$: Applying Active Concepts to Workflow Management, in: GI Datenbank Rundbrief 14 (1994) November, pp. 21-24.

Kappel et al. (1995)

Kappel, G., Rausch-Schott, S., Retschitzegger, W., Vieweg, S., TriGS$_{flow}$ - Active Object-Oriented Workflow Management, in: Proceedings of the 28th Hawaii Conference on Systems Science, 1995.

Kappel/Schrefl (1989)

Kappel, G., Schrefl, M., A Behavior Integrated Entity-Relationship Approach for the Design of Object-Oriented Databases, in: C. Batini (ed.), Entity-Relationship Approach: A Bridge to the User, Proceedings of the 7th International Conference on Entity-Relationship Approach, Amsterdam et al.: North-Holland 1989, pp. 101-122.

Kappel/Schrefl (1991)

Kappel, G., Schrefl, M., Object/Behavior Diagrams, in: Proceedings Seventh International Conference on Data Engineering, IEEE Computer Society Press, Los Alamitos 1991, pp. 530-539.

Karadimce/Urban (1991)

Karadimce, A.P., Urban, S.D., Diagnosing Anomalous Rule Behaviour in Databases with Integrity Maintenance Production Rules, in: J. Göers, A. Heuer, G. Saake (Eds.), Proceedings of the Third Workshop on Foundations of Models and Languages for Data and Objects, Technical University Clausthal 1991, pp. 77-102.

Karagiannis (1994)

Karagiannis, D., Wissensbasierte Datenbanken, Handbuch der Informatik Band 8.2, München: R. Oldenbourg 1994.

Keen/Lakos (1994)

Keen, C., Lakos, C., Information Systems Modelling Using LOOPN++, An Object Petri Net Shceme, in: A. Verbraeck, H.G. Sol, P.W.G. Bots (Eds.), Proceedings of the 4th International Working Conference on Dynamic Modelling and Information Systems, Delft: Delft University Press 1994, pp. 31-52.

Keller (1994)

Keller, G., Modellierung von Geschäftsprozessen mit der EPK-Methodik, in: Informationssystem Architekturen; Wirtschaftsinformatik-Rundbrief des GI-Fachausschusses 5.2, 1 (1994) 2.

Kendler (1982)
Kendler, H., Triggerkonzept, in: Informatik-Spektrum 5 (1982) 5, pp. 255.

Kent (1983)
Kent, W., Fact-Based Data Analysis and Design, in: C.G. Davis, S. Jajodia, P.-A. Ng, R.T. Yeh (Eds.), Proceedings of the Third International Conference on Entity-Relationship Approach to Software Engineering, Amsterdam et al.: Elsevier 1983, pp. 3-53.

Khoshafian (1990)
Khoshafian, S., Insight into Object-Oriented Databases, in: Information and Software Technology 32 (1990) 4, pp. 274-289.

Khoshafian et al. (1992)
Khoshafian, S., Chan, A., Wong, A., Wong, H.K.T., A guide to developing client/server SQL applications, San Matheo: Morgan Kaufmann Publishers 1992.

Kiernan et al. (1990)
Kiernan, J., de Maindreville, C., Simon, E., Making deductive databases a practical technology: A step forward, in: Proceeding of the ACM SIGMOD Conference on Management of Data, 1990, pp. 237-246.

Kieser/Kubicek (1978)
Kieser, A., Kubicek, H., Organisationstheorien II, Stuttgart et al.: Kohlhammer 1978.

Kieser/Kubicek (1992)
Kieser, A., Kubicek, H., Organisation, 3rd Ed., Berling/New York: De Gruyter 1992.

Kim et al. (1992)
Kim, W., Lee, Y.-J., Seo, J., A Framework for Supporting Triggers in Object-Oriented Database Systems, in: International Journal of Intelligent and Cooperative IS 1 (1992) 1, pp. 127-143.

Knolmayer et al. (1994)
Knolmayer, G., Herbst, H., Schlesinger, M., Enforcing Business Rules by the Application of Trigger Concepts, in: Swiss National Science Foundation (Ed.), Proceedings Priority Programme Informatics Research, Information Conference Module 1, Bern: Swiss National Science Foundation 1994, pp. 24-30.

Knolmayer/Herbst (1993)
Knolmayer, G., Herbst, H., Business Rules, in: Wirtschaftsinformatik 35 (1993) 4, pp. 386-390.

Knolmayer/Myrach (1990)
Knolmayer, G., Myrach, T., Anforderungen an Tools zur Darstellung und Analyse von Datenmodellen, in: Handbuch der modernen Datenverarbeitung 27 (1990) 152, pp. 90-102.

Knolmayer/Schlesinger (1994a)
Knolmayer, G., Schlesinger, M., Die Transformierbarkeit von Applikationstriggern in Datenbanktrigger am Beispiel der Oracle-Entwicklungsumgebung und des BALICO-Systems, Working Paper 49, Institute of Information Systems, University of Bern 1994.

Knolmayer/Schlesinger (1994b)
Knolmayer, G., Schlesinger, M., Geschäftsregeln in einem System der Liegenschaftsverwaltung und ihre Abbildbarkeit in einem kommerziell verfügbaren Datenbanksystem, in: GI Datenbank Rundbrief 14 (1994) November, pp. 28-31.

Knolmayer/Spahni (1993)

Knolmayer, G., Spahni, D., Darstellung und Vergleich ausgewählter Methoden zur Bestimmung von IS-Architekturen, in: H. Reichel (Ed.), Proceedings of the GI-Jahrestagung "Informatik - Wirtschaft - Gesellschaft", Berlin et al.: Springer 1993, pp. 101-104.

Kosciuszko (1992a)

Kosciuszko, E., Putting Confidence in Your Integrity, in: Database Programming & Design 5 (1992) 6, pp. 46-51.

Kosciuszko (1992b)

Kosciuszko, E., How to implement integrity in Oracle, part II, in: Database Programming & Design 5 (1992) 7, pp. 40-45.

Kotz et al. (1988)

Kotz, A. M., Dittrich, K.R., Mülle, J.A., Supporting Semantic Rules by a Generalized Event/Trigger Mechanism, in: J.W. Schmidt, S. Ceri, M. Missikoff (Eds.), Proceedings of the International Conference on Extending Database Technology, Berlin et al.: Springer 1988, pp. 76-91.

Kotz Dittrich (1992)

Kotz Dittrich, A. M., Adding Active Functionality to an Object-Oriented Database System - A Layered Approach, in: K.R. Dittrich (Ed.), Proceedings of the Workshop on Object-Oriented Database Systems at Work, SI 1992, pp. 1-20.

Küng (1994a)

Küng, P., Datenbanksysteme: Entwicklungsstand, Anforderungen und Bedeutung neuerer Konzepte, Zürich: Verlag der Fachvereine 1994.

Küng (1994b)

Küng, P., O2 und Objectstore vor Objectivity und Ontos: Zehn objektorientierte DBMS im Vergleich, in: Output 23 (1994) 6, pp. 60-63.

Kung/Sølvberg (1986)

Kung, C.H., Sølvberg, A., Activity Modeling and Behaviour Modeling, in: T.W. Olle, H.G. Sol, A.A. Verijn-Stuart (Eds.), Information Systems Design Methodologies: Improving the Practice, Amsterdam et al.: Elsevier 1986, pp. 145-171.

Lafue (1982)

Lafue, G.M.E., Semantic Integrity Dependencies And Delayed Integrity Checking, in: Proceedings of the Eighth International Conference on Very Large Databases, Los Alamitos: IEEE Computer Society Press 1982, pp. 292-299.

Laguna Beach Participants (1989)

Laguna Beach Participants, Future Directions in DBMS Research, in: SIGMOD RECORD 18 (1989) 1, pp. 17-26.

Lamb et al. (1991)

Lamb, C., Landis, G., Orenstein, J., Weinreb, D., The ObjectStore Database System, in: Communications of the ACM 34 (1991) 10, pp. 50-63.

Lazarevic/Misic (1991)

Lazarevic, B., Misic, V., Extending the entity-relationsship model to capture dynamic behaviour, in: European Journal of Information Systems 1 (1991) 2, pp. 95-106.

Le Blanc/Korn (1994)

Le Blanc, L.A., Korn, W.M., A phased approach to the evaluation and selection of CASE tools, in: Information and Software Technology 36 (1994) 5, pp. 267-273.

Lee et al. (1994)
Lee, R.M., Bons, R.W.H., Wrigley, C.D., Wagenaar, R.W., Automated Design of Electronic Trade Procedures Using Documentary Petri Nets, in: A. Verbraeck, H.G. Sol, P.W.G. Bots (Eds.), Proceedings of the Fourth International Working Conference on Dynamic Modelling and Information Systems, Delft: Delft University Press 1994, pp. 137-150.

Leikauf (1989)
Leikauf, P., Konsistenzsicherung durch Verwaltung von Inkonsistenzen, in: T. Härder (Ed.), Proceedings of the GI/SI-Fachtagung Datenbanksysteme in Büro, Technik und Wissenschaft, Berlin et al.: Springer 1989, pp. 135-153.

Leikauf (1990)
Leikauf, P., Konsistenzsicherung durch Verwaltung von Inkonsistenzen, Dissertation Nr. 9208, Zürich: ETH Zürich 1990.

Lenzerini/Santucci (1983)
Lenzerini, M., Santucci, G., Cardinality Constraints in the Entity-Relationship Approach, in: C.G. Davis, S. Jajodia, P.-A. Ng, R.T. Yeh (Eds.), Proceedings of the Third International Conference on Entity-Relationship Approach to Software Engineering, Amsterdam et al.: Elsevier 1983, pp. 529-549.

Liebowitz (1988)
Liebowitz, J. (Ed.), Expert Systems Application to Telecommunications, New York et al.: John Wiley & Sons 1988.

Likert (1967)
Likert, R., The Human Organization, New York: McGraw-Hill 1967.

Lingat et al. (1987)
Lingat, J.Y., Nobecourt, P., Rolland, C., Behaviour Management in Database Applications, in: P.M. Stocker, W. Kent, P. Hammersley (Eds.), Proceedings of the Thirteenth International Conference on Very Large Databases, Los Alamitos: IEEE Computer Society Press 1987, pp. 185-196.

Lipeck (1989)
Lipeck, U. W., Zur dynamischen Integrität von Datenbanken: Grundlagen der Spezifikation und Überwachung, Berlin et al.: Springer 1989.

Lipeck (1992)
Lipeck, U.W., Integritätszentrierter Datenbank-Entwurf, in: EMISA-Forum (1992) 2, pp. 41-55.

Lockemann et al. (1990)
Lockemann, P.C., Kemper, A., Moerkotte, G., Future Database Technology: Driving Forces and Directions, in: A. Blaser (Ed.), Proceedings of the International Symposium on Database Systems of the 90s, Berlin et al.: Springer 1990, pp. 15-33.

Lockemann/Radermacher (1990)
Lockemann, P.C., Radermacher, K., Konzepte, Methoden und Modelle zur Datenmodellierung, in: Handbuch der modernen Datenverarbeitung 27 (1990) 152, pp. 3-16.

Lockemann/Walter (1993)
Lockemann, P.C., Walter, H.D., Activities in Object Bases, in: W. Paton (Ed.), Proceedings of the 1st International Workshop on Rules in Database Systems, Berlin et al.: Springer 1993, pp. 3-22.

Losavio et al. (1994)

    Losavio, F., Matteo, A., Schlienger, F., Object-oriented methodologies of Coad and Yourdon and Booch: comparison of graphical notations, in: Information and Software Technology 36 (1994) 8, pp. 503-514.

Loucopoulos (1992)

    Loucopoulos, P., Conceptual Modeling, in: P. Loucopoulos (Ed.), Conceptual Modeling, Databases, and CASE, New York et al.: John Wiley & Sons 1992, pp. 1-26.

Loucopoulos et al. (1990)

    Loucopoulos, P., McBrien, P., Persson, U., Schumacker, F., Vasey, P., TEMPORA - Integrating Database Technology. Rule Based Systems and Temporal Reasoning für Effective Software, in: Proceedings of the ESPRIT '90 Conference, 1990, pp. 388-411.

Manthey (1993)

    Manthey, R., Rules and Objects - Issues In The Design And Development Of DOOD Systems, EDBT Summer School, 1993.

March/Simon (1958)

    March, J.G., Simon, H.A., Organizations, New York et al.: John Wiley & Sons 1958.

Mark (1983)

    Mark, L., What is the Binary Relationship Approach, in: C.G. Davis, S. Jajodia, P.-A. Ng, R.T. Yeh (Eds.), Proceedings of the Third International Conference on Entity-Relationship Approach to Software Engineering, Amsterdam et al.: Elsevier 1983, pp. 205-220.

Marti (1983)

    Marti, R.W., Integrating Database and Program Descriptions using an ER-Data Dictionary, in: C.G. Davis, S. Jajodia, P.-A. Ng, R.T. Yeh (Eds.), Proceedings of the Third International Conference on Entity-Relationship Approach to Software Engineering, Amsterdam et al.: Elsevier 1983, pp. 377-392.

Martin (1985)

    Martin, J., Fourth-Generation Languages, Vol. 1, Englewood Cliffs: Prentice-Hall 1985.

Martin/Leben (1986)

    Martin, J., Leben, J., Fourth-Generation Languages, Vol. 2, Englewood Cliffs: Prentice-Hall 1986.

Martin (1990)

    Martin, J., Information Engineering II; Planning and Analysis, Englewood Cliffs: Prentice Hall 1990.

Martin/Odell (1992)

    Martin, J., Odell, J., Object-Oriented Analysis & Design, Englewood Cliffs: Prentice-Hall 1992.

McBrien et al. (1991)

    McBrien, P., Niézette, M., Pantazis, D., Seltveit, A.H., Sundin, U., Theodoulidis, B., Tziallas, G., Wohed, R., A Rule Language to Capture and Model Business Policy Specifications, in: R. Anderson, J.A. Bubenko jr., A. Sølvberg (Eds.), Proceedings of the Third Nordic Conference on Advanced Information Systems Engineering (CAiSE '91), Berlin et al.: Springer 1991, pp. 307-318.

McCarthy/Dayal (1989)

McCarthy, D.R., Dayal, U., The Architecture Of An Active Database Management System, in: SIGMOD RECORD 18 (1989) 2, pp. 215-224.

McDermid (1990)

McDermid, D., The state dependency diagram, in: Software Engineering Journal (1990) 5, pp. 165-173.

McGregor (1960)

McGregor, D., The human side of Enterprise, New York: McGraw-Hill 1960.

McMenamin/Palmer (1984)

McMenamin, S.M., Palmer, J.F., Essential Systems Analysis, Englewood Cliffs: Prentice-Hall 1984.

Melton/Simon (1993)

Melton, J., Simon, A.R., Understanding the New SQL: A Complete Guide, San Mateo: Morgan Kaufmann 1993.

Melton (1993)

Melton, J., The Structure of SQL3, in: Database Programming & Design 6 (1993) 7, pp. 65-67.

Mertens/Hofmann (1986)

Mertens, P., Hofmann, J., Aktionsorientierte Datenverarbeitung, in: Informatik-Spektrum 9 (1986) 9, pp. 323-333.

Meseguer (1990)

Meseguer, P., A New Method to Checking Rule Bases for Inconsistency: A Petri Net Approach, in: L.C. Aiello, E. Sandewall, G. Hagert and B. Gustavsson (eds.), Proceedings of the 9th European Conference on Artificial Intelligence 1990, pp. 437-442.

Mintzberg (1979)

The Structuring of Organizations, Englewood Cliffs: Prentice-Hall 1979.

Misra/Jalics (1988)

Misra, S.K., Jalics, P.J., Third-Generation versus Fourth-Generation Software Development in: IEEE Software, (1988) July, pp. 8-14.

Morgenstern (1983)

Morgenstern, A., Active Databases as a Paradigm for Enhanced Computing Environments, in: M. Schkolnick, C. Thanos (Eds.), Proceedings of the 9th International Conference on Very Large Databases, 1983, pp. 34-42.

Morgan (1993)

Morgan, G., Imaginization, Newbury Park 1993.

Moriarty (1993a)

Moriarty, T., The Next Paradigm, in: Database Programming & Design 6 (1993) 2, pp. 66-69.

Moriarty (1993b)

Moriarty, T., Business Rule Analysis, in: Database Programming & Design 6 (1993) 4, pp. 66-69.

Morris/Brandon (1993)

Morris, D., Brandon, J., Re-engineering Your Business, New York et al.: McGraw-Hill 1993.

Mück (1994)
Mück, T.A., Design Support for Initiatives and Policies in Conceptual Models of Information Systems - A Statechart Approach, in: J.F. Nunamaker and R.H. Sprague (Eds.), Proceedings of the 27th Annual Hawaii International Conference on System Sciences, Vol. IV, IEEE Computer Society Press, Los Alamitos 1994, pp. 743-752.

Müller-Nobiling (1980)
Müller-Nobiling, H.-M, Organisationshandbuch, in: E. Grochla (Ed.), Handwörterbuch der Organisation, 2nd Ed., Stuttgart: Poeschel 1980, col. 187-200.

Murray/Murray (1988)
Murray, J.T., Murray, M.J., Expert Systems in Data Processing - A Professional Guide, New York et al.: McGraw-Hill 1988.

Myrach (1995)
Myrach, T., Konzeption und Stand des Einsatzes von Data Dictionaries, Heidelberg: Physica 1995.

Myrach et al. (1996)
Myrach, T., Knolmayer, G.F., Barnert, R., On Ensuring Keys and Referential Integrity in the Temporal Database Language TSQL2, in: H.-M. Haav and B. Thalheim (Eds.), Proceedings of the Second International Baltic Workshop on Databases and Information Systems, Volume 1, 1996, pp. 171-181.

Naqvi/Ibrahim (1993a)
Naqvi, W., Ibrahim, M.T., Rule and Knowledge Management in an Active Database System, in: W. Paton (Ed.), Proceedings of the 1st International Workshop on Rules in Database Systems, Berlin et al.: Springer 1993, pp. 58-69.

Naqvi/Ibrahim (1993b)
Naqvi, W., Ibrahim, M.T., REFLEX Active Database Model: Application of Petri-Nets, in: V. Marík, J. Lazanský, R.R. Wagner (Eds.), Proceedings of the Fourth International Conference on Database and Expert Systems Applications, Berlin et al.: Springer 1993, pp. 233-240.

Navathe (1992)
Navathe, S.B., The Next Ten Years of Modeling, Methodologies, and Tools, in: G. Pernul, A.M. Tjoa (Eds.), Proceedings of the 11th International Conference on the Entity-Relationship Approach, Berlin et al.: Springer 1992, pp. 4-6.

Nazareth (1993)
Nazareth, D.L., Investigating the Applicability of Petri Nets for Rule-Based System Verification, in: IEEE Transactions on Knowledge and Data Engineering 4 (1993) 3, pp. 402-415.

Neuhold (1983)
Neuhold, E.J., Development Methods for Event and Message Based Application Systems, in: E. Knuth, E.J. Neuhold (Eds.), Proceedings of the Conferene on Operating Systems - Specification and Design of Software Systems, Berlin et al.: Springer 1983, pp. 28-55.

Nicolas/Yazdanian (1978)
Nicolas, J.M., Yazdanian, K., Integrity Checking in Deductive Databases, in: I.H. Gallaire, J. Minker (Eds.), Proceedings of the Symposium on Logic and Data Bases, New York/London: Plenum Press 1978, pp. 325-344.

Nijssen/Halpin (1989)
Nijssen, G.M., Halpin, T.A., Conceptual Schema and Relational Database Design - A Fact Oriented Approach, Englewood Cliffs: Prentice-Hall 1989.

Informix (1993)
Informix, Trigger und Stored Procedures in Informix, in: Informix Times (1993) 2, pp. 10-11.

ObjectDesign (1991)
Object Design, ObjectStore Technical Overview, Release 1.0, 1991.

Olle et al. (1988)
Olle, T.W., Hagelstein, J., Macdonald, I.G., Rolland, C., Sol, H.G., Van Assche, F., Verijn-Stuart, A.A., Information Systems Methodologies, A Framework for Understanding, Wokingham et al.: Addison-Wesley 1988.

Österle/Gutzwiller (1992a)
Österle, H., Gutzwiller, T., Konzepte angewandeter Analyse- und Design-Methoden - Band 1: Ein Referenz-Metamodell für die Analyse und das System-Design, Hallbergmoos: AIT 1992.

Österle/Gutzwiller (1992b)
Österle, H., Gutzwiller, T., Konzepte angewandeter Analyse- und Design-Methoden - Band 2: Beispiel, Hallbergmoos: AIT 1992.

Page-Jones (1990)
Page-Jones, M., The Practical Guide to Structured Systems Design, Englewood Cliffs: Prentice-Hall 1990.

Perrow (1970)
Perrow, C., Organizational Analysis: A Sociological Review, London 1970.

Perry/Denna (1995)
Perry, L.T., Denna, E.L., Re-engineering Redux: An Agenda for Next-Generation BPR, in: Business Change & Re-engineering 2 (1995) 3, pp. 61-71.

Petri (1962)
Petri, C.A., Komunikation mit Automaten, Dissertation, Bonn: Institut für Angewandte Mathematik der Universität Bonn 1962.

Pieper (1992)
Pieper, F., 3GL versus 4GL: Ein Praxisvergleich, in Datenbank FOKUS (1992) 7/8, pp. 20-27.

Plotkin (1995)
Plotkin, D., Taking the Repository Plunge, in: Database Programming & Design 8 (1995) 2, pp. 67-69.

Pohl (1981)
Pohl, H.-J., Grundlagen der Organisationsforschung - Die Gestaltung von Organisationsstrukturen und ihre empirischer Erforschung, Bremen 1981.

Pohl (1993)
Pohl, K., The Three Dimensions of Requirements Engineering, in: C. Rolland, F. Bodart, C. Cauvet (Eds.), Proceedings of the Fifth International Conference on Advanced Information Systems Engineering, Berlin et al.: Springer 1993, pp. 275-292.

Poo (1992)
Poo, C.-C. D., A Framework for Software Maintenance, in: P. Loucopoulos (Ed.), Proceedings of the Fourth International Conference on Advanced Information Systems Engineering (CAiSE '92), Berlin et al.: Springer 1992, pp. 88-104.

Probst (1987)
Probst, G., Ordnungsprozesse in sozialen Systemen aus ganzheitlicher Sicht, Berlin et al. Springer 1987.

Pugh et al. (1969)

Pugh, D.S., Hickson, D.J., Hinings, C.R., An Empirical Taxonomy of Structures of Work, in: Administrative Science Quarterly 14 (1969), pp. 115-126.

Quang/Chartier-Kastler (1991)

Quang, P.T., Chartier-Kastler, C., Merise in Practice, Macmillan Education 1991.

Quer/Olivé (1993)

Quer, C., Olivé, A., Object Integration in Object-Oriented Deductive Conceptual Models, in: C. Rolland, F. Bodart, C. Cauvet (Eds.), Proceedings of the 5th International Conference on Advanced Information Systems Engineering (CAiSE '93), Berlin et al.: Springer 1993, pp. 374-396.

R&O (1994a)

R&O Softwaretechnik, Rochade 4.1x - Referenzhandbuch, Germering: R&O GmbH 1994.

R&O (1994b)

R&O Softwaretechnik, Autopilot - Administrationshandbuch, Germering: R&O GmbH 1994.

Ramakrishnan et al. (1993)

Ramakrishnan, R., Roth, W.G., Seshadri, P., Srivastava, D., Sudarshan, S., The CORAL Deductive Database System, in: SIGMOD RECORD 22 (1993) 2, pp. 544-545.

Rebsamen (1983)

Rebsamen, J., Datenbankentwurf im Dialog - Integrierte Beschreibung von Strukturen, Transaktionen und Konsistenz, Dissertation, Zürich: ETH Zürich 1983.

Reinert (1994)

Reinert, J., Eventbegriffe in Datenbanksystemen, in: GI Datenbank Rundbrief 14 (1994) November, pp. 11-14.

Rennhackkamp (1992)

Rennhackkamp, M., Using Ingres Event Alerters, in: DBMS 5 (1992) 11, pp. 86-108.

Reuter (1987)

Reuter, A., Massnahmen zur Wahrung von Sicherheits- und Integritätsbedingungen, in: P.C. Lockemann, J.W. Schmidt (Eds.), Datenbank-Handbuch, Berlin et al.: Springer 1987, pp. 379-409.

Richter (1984a)

Richter, G., Netzmodelle für die Bürokommunikation - Teil 1, in: Informatik-Spektrum 7 (1984) 6, pp. 210-220.

Richter (1984b)

Richter, G., Netzmodelle für die Bürokommunikation - Teil 2, in: Informatik-Spektrum 7 (1984) 7, pp. 28-40.

Robson/Henderson (1993)

Robson, A.P., Henderson, W.D., Data Flow Model for the implementation of real-time programs on simple target machines, in: Information and Software Technology 35 (1993) 1, pp. 3-16.

Rochfeld/Tardieu (1983)

Rochfeld, A., Tardieu, H., MERISE: An Information System Design and Development Methodology, in: Information & Management, 6 (1983), pp. 143-159.

Rockart/Short (1991)
Rockart, J.F., Short, J.E., The Networked Organization and the Management of Interdependence, in: M.S.S. Morton (Ed.), The corporation of the 1990s, New York/Oxford: Oxford University 1991, pp. 189-219.

Roethlisberger/Dickson (1939)
Roethlisberger, F.J., Dickson, W.J., Management and the Worker: An Account on a Research Program Conducted by the Western Electric Company, Hawthorne Works, Harvard University Press 1939.

Rolland/Cauvet (1992)
Rolland, C., Cauvet, C., Trends and Perspectives in Conceptual Modeling, in: P. Loucopoulos (Ed.), Conceptual Modeling, Databases, and CASE, New York et al.: John Wiley & Sons 1992, pp. 27-48.

Rolland/Proix (1992)
Rolland, C., Proix, C., A Natural Language Approach for Requirements Engineering, in: P. Loucopoulos (Ed.), Proceedings of the Fourth International Conference on Advanced Information Systems Engineering (CAiSE '92), Berlin et al.: Springer 1992, pp. 257-277.

Rolland/Richard (1982)
Rolland, C., Richard, C., The REMORA Methodology for Information Systems Design and Management, in: T.W. Olle, H.G. Sol, A.A. Verijn-Stuart (Eds.), Information Systems Design Methodologies: A Comparative Review, Amsterdam et al.: North-Holland 1982, pp. 369-426.

Rolland et al. (1988)
Rolland, C., Foucaut, O., Benci, G., Conception des systèmes d'information, La méthode REMORA, Paris: Eyrolles 1988.

Romberg (1995)
Romberg, E.A., Meta-Entities: Keeping Pace With Change, in: Database Programming & Design 8 (1995) 1, pp. 54-59.

Rosenstengel/Winand (1982)
Rosenstengel, B., Winand, U., Petri-Netze, eine anwendungsorientierte Einführung, Braunschweig/Wiesbaden: Vieweg 1982.

Rückle/Terhart (1986)
Rückle, D., Terhart, K., Die Befolgung von Umweltschutzauflagen als betriebswirt-schaftliches Entscheidungsproblem, in: Schmalenbachs Zeitschrift für betriebswirtschaftliche Forschung 38 (1986) 5, pp. 393-424.

Rumbaugh et al. (1991)
Rumbaugh, J., Blaha, M., Premerlani, W., Eddy, F., Lorensen, W., Object Oriented Modeling and Design, Englewood Cliffs: Prentice-Hall 1991.

Rusinkiewicz/Sheth (1995)
Rusinkiewicz, M., Sheth, A., Specification and Execution of Transactional Workflows, in: W. Kim (Ed.), Modern Database Systems: The Object Model, Interoperability, & Beyond, New York: ACM Press 1995, pp. 592-620.

Rytz (1994)
Rytz, M., Stand der Nutzung von Datenbank-Triggern in der Anwendungsentwick-lung - Eine empirische Untersuchung zum Einsatz von Triggerkonzepten in Oracle, Sybase und Ingres, Masters Thesis, Institute of Information Systems, University of Bern 1994.

Sakai (1983)
 Sakai, H., A Method for Entity-Relationship Behavior Modeling, in: C.G. Davis, S. Jajodia, P.-A. Ng, R.T. Yeh (Eds.), Proceedings of the Third International Conference on Entity-Relationship Approach to Software Engineering, Amsterdam et al.: Elsevier 1983, pp. 111-129.

Sakai (1990)
 Sakai, H., An Object Behavior Modeling Method, in: A.M. Tjoa, R. Wagner (Eds.), Proceedings of the International Conference on Database and Expert Systems Applications, Wien/New York: Springer 1990, pp. 42-48.

Sakthivel/Moily (1993)
 Sakthivel, S., Moily, J.P., Analytical verification of information system requirements using Petri net properties, in: Information and Software Technology 35 (1993) 2, pp. 89-100.

Sandifer/Von Halle (1991)
 Sandifer, A., Von Halle, B., Linking Rules to Models, in: Database Programming & Design 4 (1991) 3, pp. 13-16.

Scheer (1991)
 Scheer, A.-W., Architektur integrierter Informationssysteme, Berlin et al.: Springer 1991.

Scheer (1995)
 Scheer, A.-W., Wirtschaftsinformatik - Referenzmodelle für industrielle Geschäftsprozesse 6th Ed., Berlin et al.: Springer 1995.

Schlesinger (1995)
 Schlesinger, M., Vergleich aktiver Mechanismen in Ingres V6.4, Oracle V7.0 und Sybase V10.0, in: H.-J. Scheibl (Ed.), Softwareentwicklung - Methoden, Werkzeuge, Erfahrungen '95, Technische Akademie Esslingen 1995, pp. 41-53.

Schlesinger/Achermann (1995)
 Schlesinger, M., Achermann, F., Vergleichende Gegenüberstellung der Trigger-Mechanismen von Ingres, Oracle und Sybase, Working Paper No. 62, Institute of Information Systems, University of Bern 1995.

Schmidt (1985)
 Schmidt, G., Grundlagen der Aufbauorganisation, Giessen: Dr. Georg Schmidt 1985.

Schmidt (1989)
 Schmidt, G., Methode und Techniken der Organisation, Giessen: Dr. Georg Schmidt 1989.

Schrefl (1990)
 Schrefl, M., Behavior Modeling by Stepwise Refining Behavior Diagrams, in: H. Kangassalo (Ed.), Proceedings of the Nineth International Conference on Entity-Relationship Approach, San Matheo: Morgan Kaufmann 1990, pp. 113-128.

Schrefl/Stumptner (1995)
 Schrefl, M., Stumptner, M., Behavior Consistent Extension of Object Life Cycles, in: M. P. Papazoglou (Ed.), 14th International Conference on Object-Oriented and Entity-Relationship Modelling, Berlin et al.: Springer 1995, 133-145.

Sellis (1995)
 Sellis, T., (Ed.), Rules in Database Systems, Berlin et al.: Springer 1995.

Shepherd/Kerschberg (1986)
Shepherd, A., Kerschberg, L., Constraint Management in Expert Database Systems, in: L. Kerschberg (Eds.), Expert Database Systems, Redwood City et al.: Benjamin/Cummings 1986, pp. 309-331.

Silverman (1987)
Silverman, B.G., Should a Manager „Hire" an Expert System? in: B.G. Silverman (Ed.), Expert systems for Business, Reading et al.: Addison-Wesley 1987.

Simon et al. (1992)
Simon, E., Kiernan, J., de Maindreville, C., Implementing high level active rules on top of a relational DBMS, in: Proceedings of the 18th International Conference on Very Large Databases, 1992, pp. 315-326.

Sinz (1990)
Sinz, E.J., Das Entity-Relationship-Modell (ERM) und seine Erweiterungen, in: Handbuch der modernen Datenverarbeitung 27 (1990) 152, pp. 17-29.

Skidmore et al. (1992)
Skidmore, S., Farmer, R., Mills, G., SSADM Version 4 - Models and Methods, Manchester, Oxford: NCC Blackwell 1992.

Smith/Smith (1977)
Smith, J.M., Smith D.C.P., Database Abstractions: Aggregation and Generalization, in: ACM Transaction on Database Systems 2 (1977) 2, pp. 105-133.

Soloviev (1992)
Soloviev, V., An Overview of Three Commercial Object-Oriented Database Management Systems: ONTOS, ObjectStore, and O2, in: SIGMOD RECORD 21 (1992) 1, pp. 93-104.

Sølvberg/Kung (1993)
Sølvberg, A., Kung, D.C., Information Systems Engineering - An Introduction, Berlin et al.: Springer 1993.

Song/Forbes (1991)
Song, I.-Y., Forbes, E.A., Schema Conversion Rules between EER and NIAM Model, in: T.J. Teorey (Ed.), Proceedings of the 10th International Conference on the Entity Relationship Approach, San Matheo: Morgan Kaufmann 1991, pp. 417-444.

Souveyet/Rolland (1990)
Souveyet, C., Rolland, C., Correction of Conceptual Schemas, in: B. Steinholtz, A. Sølvberg, L. Bergman (Eds.), Proceedings of the Second Nordic Conference on Advanced Information Systems Engineering (CAiSE '90), Berlin et al.: Springer 1990, pp. 152-174.

Spurr (1989)
Spurr, K., Introduction to the ISO IRDS Standards, in: S. Holloway (Ed.), The Future of Data Dictionaries, Brookfield: Aldershot 1989, pp. 7-18.

Stefik et al. (1983)
Stefik, M., Aikins, J., Balzer, R., Benoit, J., Birnbaum, L., Hayes-Roth, F., Sacerdoti, E., Basic Concepts for Building Expert Systems, in: F. Hayes-Roth, D.A. Waterman, D.B. Lenat (Eds.), Building Expert Systems, Reading et al.: Addison-Wesley 1983, p. 59-87.

Steinbauer (1983)
Steinbauer, D., Transaktionen als Grundlage zur Strukturierung und Integritätssicherung in Datenbank-Anwendungssystemen, PhD Dissertation, Betriebswirtschaftliches Institut, University of Erlangen-Nürnberg 1983.

Steinbauer/Wedekind (1985)

Steinbauer, D., Wedekind, H., Integritätsaspekte in Datenbanksystemen, in: Informatik-Spektrum 8 (1985) 2, pp. 60-68.

Stonebraker (1992)

Stonebraker, M., The Integration of Rule Systems and Databases, in: IEEE Transactions on Knowledge and Data Engineering 4 (1992) 5, pp. 415-423.

Stonebraker et al. (1987)

Stonebraker, M., Hanson, E., Hong, C.-H., The Design of the POSTGRES Rules System, in: Proceedings of the IEEE International Conference on Data Engineering 1987, pp. 365-374.

Stonebraker et al. (1989)

Stonebraker, M., Hearst, M., Potamianos, S., A Commentary on the Postgres Rules System, in: SIGMOD RECORD 18 (1989) 3, pp. 5-11.

Streng (1994)

Streng, R.J., BPR needs BIR and BTR: The PIT-framework for Business Re-engineering, in: Proceedings of the Second SISnet Conference, University of Navarra 1994.

Strong/Miller (1995)

Strong, D.M., Miller, S.M., Exceptions and Exception Handling in Computerized Information Processes, in: ACM Transactions on Information Systems 13 (1995) 2, pp. 206-233.

Suter (1995)

Suter, P., Regeln in der Organisationslehre, Masters Thesis, Institute of Information Systems, University of Bern 1995.

Takashima (1986)

Takashima, Q., Characteristics of Current Expert Systems, in: T. Bernold (Ed.), Expert Systems and Knowledge Engineering, Amsterdam et al.: North-Holland 1986, p. 31-50.

Tanaka (1992)

Tanaka, A.K., On Conceptual Design of Active Databases, PhD Thesis, Georgia Institute of Technology 1992.

Tanaka et al. (1991)

Tanaka, A.K., Navathe, S.B., Chakravarthy, S., Karlapalem, K., ER-R: An Enhanced ER Model with Situation-Action Rules to Capture Application Semantics, in: T.J. Teorey (Ed.), Proceedings of the 10th International Conference on the Entity Relationship Approach, San Matheo: Morgan Kaufmann 1991, pp. 59-75.

Tari (1992)

Tari, Z., On the Design of Object-Oriented Databases, in: G. Pernul, A.M. Tjoa (Eds.), Proceedings of the 11th International Conference on the Entity-Relationship Approach, Berlin et al.: Springer 1992, pp. 389-405.

Taylor (1911)

Taylor, F., The principles of scientific management, New York: Harper & Row 1911.

Teisseire et al. (1994)

Teisseire, M., Poncelet, P., Cicchetti, R., Dynamic Modelling with Events, in: G. Wijer, S. Brinkkemper, T. Wasserman (Eds.), Proceedings of the 6th Conference on Advanced Information Systems Engineering (CAiSE '94), Berlin et al.: Springer 1994, pp. 186-199.

Thalheim (1992)
Thalheim, B., An Overview on Database Theory, in: GI Datenbank Rundbrief 10 (1992) November, pp. 2-13.

Thaller (1994)
Thaller, G.E., Verifikation und Validation, Braunschweig/Wiesbaden: Vieweg 1994.

Theisges/Denk (1991)
Theisges, W., Denk, H., Zustandsübergangsdiagramme als strukturbestimmende Sicht eines Systems, in: M. Timm (Ed.), Requirements Engineering '91 - 'Structured Analysis' und verwandte Ansätze, Berlin et al.: Springer 1991, pp. 93-129.

Theodoulidis et al. (1994)
Theodoulidis, C., Wangler, B., Loucopoulos P., Requirements Specification in TEMPORA, in: B. Steinholtz, A. Sølvberg, L. Bergman (Eds.), Proceedings of the Second Nordic Conference on Advanced Information Systems Engineering (CAiSE '90), Berlin et al.: Springer 1994, pp. 264-282.

Thompson (1967)
Thompson, J.D., Organizations in action, New York: McGraw-Hill 1967.

Tsalgatidou et al. (1990)
Tsalgatidou, A., Karakostas, V., Loucopoulos, P., Rule-Based Requirements Specification and Validation, in: B. Steinholtz, A. Sølvberg, L. Bergman (Eds.), Proceedings of the Second Nordic Conference on Advanced Information Systems Engineering (CAiSE '90), Berlin et al.: Springer 1990, pp. 251-263.

Tsalgatidou et al. (1994)
Tsalgatidou, A., Gouscos, D., Halatsis, C., Dynamic Process Modelling Through Multi-Level RBNs, in: A. Verbraeck, H.G. Sol, P.W.G. Bots (Eds.), Proceedings of the Fourth International Conference on Dynamic Modeling and Information Systems, Delft: Delft University Press 1994, pp. 327-341.

Ullman (1989)
Ullman, J.D., Principles of Database and Knowledge Base Systems, Volumes I and II, Rockville: Computer Science Press 1989.

Urban et al. (1992)
Urban, S.D., Karadimce, A.P., Nannapaneni, R.B., The Implementation and Evaluation of Integrity Maintenance Rules in an Object-Oriented Database, in: IEEE Computer Society (ed.), Proceedings Eighth International Conference on Data Engineering, IEEE Computer Society Press, Los Alamitos 1992, pp. 565-572.

Urban et al. (1994)
Urban, S.D., Shah, J.J., Rogers, M., Jeon, D.K., Ravi, P., Bliznakov, P., A heterogeneous, active database architecture for engineering data management, in: International Journal of Computer Integrated Manufacturing 7 (1994) 5, pp. 276-293.

Van Assche et al. (1988)
Van Assche, F., Layzell, P., Loucopoulos, P., Speltincx, G., Information systems development: a rule-based approach, in: Journal of Knowledge-Based Systems 1 (1988) 4, pp. 227-234.

Venkatraman (1991)
Venkatraman, N., IT-Induced Business Reconfiguration, in: M.S.S. Morton (Ed.), The corporation of the 1990s, New York/Oxford: Oxford University 1991, p. 122-158.

Von Halle (1993a)
> Von Halle, B., Business Rules in Practice, in: Database Programming & Design 6 (1993) 7, pp. 15-18.

Weber (1972)
> Weber, M., Wirtschaft und Gesellschaft, 5th Edition, Tübingen 1972.

Weber et al. (1983)
> Weber, W., Stucky, W., Karszt, J, Integrity checking in data base systems, in: Information Systems 8 (1983) 2, pp. 125-136.

Wedekind(1976)
> Wedekind, H., Die Überprüfung von semantischen Integritätsbedingungen in Relationalen Datenbanksystemen, in: E. J. Neuhold (Ed.), Proceeding of the Sixth GI Jahrestagung, Berlin et al.: Springer 1976, pp. 282-300.

Wedekind (1983)
> Wedekind, H., Datenintegrität der Anwendung aus wissenschaftstheoretischer Sicht, in: Elektronische Rechenanlagen 25 (1983) 6, pp. 133-142.

Widom (1993)
> Widom, J., Deductive and Active Databases: Two Paradigms or Ends of a Spectrum, in: W. Paton (Ed.), Proceedings of the 1st International Workshop on Rules in Database Systems, Berlin et al.: Springer 1993, pp. 306-315.

Widom/Ceri (1996)
> Widom, J., Ceri, S., Active Database Systems - Triggers and Rules. For Advanced Database Processing, San Francisco: Morgan Kaufmann 1996.

Widom et al. (1991)
> Widom, J., Cochrane, R.J., Lindsay, B.G., Implementing Set-Oriented Production Rules as an Extension to Starburst, in: G.M. Lohman, A. Sernadas, R. Camps (Eds.), Proceedings of the 17th International Conference on Very Large Databases, 1991, pp. 275-285.

Widom/Finkelstein (1990)
> Widom, J., Finkelstein, S.J., Set-Oriented Production Rules in Relational Database Systems, in: SIGMOD RECORD 19 (1990) 2, pp. 259-270.

Yourdon (1989)
> Yourdon, E., Modern Structured Analysis, Englewood Cliffs: Prentice-Hall 1989.

Zicari/Bauzer-Medeiros (1992)
> Zicari, R., Bauzer-Medeiros, C., New Generation Database Systems, in: P. Loucopoulos (Ed.), Conceptual Modeling, Databases, and CASE, New York et al.: John Wiley & Sons 1992, pp. 139-161.

APPENDIXES

Appendix A: Syntax for Business Rules

Extended Backus-Naur form (EBNF):

`::=`	is equal to
`<x>`	x is a non-terminal symbol
`x`	x is a terminal symbol
`x ¦ y`	Selection between x and y
`{x}`	x appears n times, where n ≥ 0
`[x]`	x appears at maximum once
`XYZ`	XYZ is a fixed component of the definition

Based on this notation the syntax for business rules is as follows:

Business Rule

```
<Business Rule>        ::= ON [Event] THEN [Action] ¦
                          ON [Event] IF [Condition] THEN
                          [Action] ¦
                          ON [Event] IF [Condition] THEN
                          [Action] ELSE [Action]
```

Event

```
<Event>                ::= '(' <Elementary event> ¦ <Composite
                          event> ')'

<Elementary event>     ::= <Data manipulation event> ¦
                          <Time point event> ¦
                          <User event>

<Data manipulation event>::= <Tuple Manipulation> ¦
                                <Property manipulation>

<Property manipulation>   ::= UPDATE OF <Property> ¦
                                RETRIEVAL OF <Property> ¦
                                DERIVATION OF <Property>

<Tuple Manipulation>   ::= INSERTION OF <Object> ¦ DELETION OF
                                <Object>

<Time point event>     ::= <Month>.<Day>.<Year>/hh:mm:ss
```

```
<User event>              ::= String, i.e., any meaningful
                              describtion of an event
<Composite events>        ::= <Disjunction> |
                              <Conjunction> |
                              <Selection> |
                              <Sequence> |
                              <Interval> |
                              <Periodical> |
                              <Delay>
<Disjunction>             ::= <Event> OR <Event>
<Conjunction>             ::= <Event> AND <Event>
<Selection>               ::= <m> OF '(' <Event list> ')'
<Sequence>                ::  <Event list>
  <Event list>            ::= <Event> ; <Event>
<Interval>                ::= <Event> WITHIN '(' <Event> ; <Event>
                              ')'
<Periodical>              ::= EACH <Number> <Event>
<Delay>                   ::= <Number> <Unit> AFTER <Event>
  <Number>                ::= Integer
  <Unit>                  ::= SECOND | MINUTE | HOUR | DAY | MONTH
                              | YEAR
<Month>                   ::= 01 | ... | 12
<Day>                     ::= 01 | ... | 31
<Year>                    ::= 1900 | ... | 9999
<Minute>                  ::= 01 | ... | 60
<Hour>                    ::= 01 | ... | 24
<Second>                  ::= 01 | ... | 60
<Property>                ::= <Property name> OF <Object>
<Object>                  ::= String
<Property name>           ::= String
```

Condition

```
<Condition>               ::= [NOT] '('
                              <Elementary condition> |
                              <Composite condition> ')'
<Elementary condition>    ::= <Static constraint> | <Dynamic
                              constraint>
  <Static constraint>     ::= <Set> | <Range> | <Relation>
    <Set>                 ::= <Property> IN '(' <Value list> ')'
```

<Range>	::= <Property> < | ≤ | = | ≥ | > <Property> | Value
<Relation>	::= EXISTS <Object> RELATED TO <Object>
<Dynamic constraint>	::= <Property> '(' <Value> ')' NOT TO '(' <Value list> ')'
<Value list>	::= <Value> { ; <Value> }
<Value>	::= String | Decimal | Date
<Composite condition>	::= <Condition> AND | OR <Condition>

Action

<Action>	::= <Elementary action> RAISE EVENT <Event name> [';' <action>]
<Elementary action>	::= <Data manipulation action> | <Message action> | <User action>
<Data manipulation action>	::= <Property manipulation> | <Tuple manipulation>
<Property manipulation>	::= UPDATE <Property> TO <Value>| RETRIEVE <Property> | DERIVE <Property> FROM <Expression>
<Tuple manipulation>	::= INSERT <Object> | DELETE <Object>
<Message action>	::= MESSAGE <Message> [TO <Actor name>]
<Message>	::= String
<Actor name>	::= String, i.e., any actor of the IS
<User action>	::= String, i.e., any meaningful describtion of an action

Appendix B: Document Types of *BURRO*

The repository system described in Chapter 6 contains the implementation of the document types as specified in this Appendix.

Document type *PROCESS*		
Attribute name	**Type**[1]	**Description**
Definition	Text	Unique name of the process ·
Description	Text	Description of the process
Sub-processes	Text (SR)	References to document type *PROCESS* (for the specification of a process structure)
Business Rule	Text (SR)	References to at maximum **one** document of the type *BUSINESS-RULE* (link to the business rule that is specialized by the process)
Business Rules	Text (SR)	List of references to documents of the type *BUSINESS-RULE* (link to business rules which specify the process)

Document type *BUSINESS-RULE*[2]		
Attribute name	**Type**	**Description**
Definition	Text	Unique name of the business rule
Description	Text	Description of the entity-type
Event	Text (SR)	References to **one** document of the type EVENT
Condition	Text (UR)	Text including references to documents of the types ENTITY-TYPE, RELATIONSHIP-TYPE, ATTRIBUTE
Then-Action	Text (UR)	Text including references to documents of the types ENTITY-TYPE, RELATIONSHIP-TYPE, ATTRIBUTE, and EVENT (events raised by the action)
Else-Action	Text (UR)	Text including references to documents of the types ENTITY-TYPE, RELATIONSHIP-TYPE, ATTRIBUTE, and EVENT (events raised by the action)

---

[1] (GR): user Guided Reference; (SR): Standard Reference; (UR): Unformatted Reference.

[2] The document type BUSINESS RULE contains furthermore attributes which facilitate the classification of rules.

Document type *EVENT*		
Attribute name	**Type**	**Description**
Definition	Text	Unique name of the event
Description	Text	Description of the entity-type
Content	Text (UR)	Event specification including references to documents of the types *EVENT-TYPE* (for the definition of complex events), *ENTITY-TYPE*, *RELATIONSHIP-TYPE,* and *ATTRIBUTE*

Document type *ENTITY-TYPE*		
Attribute name	**Type**	**Description**
Definition	Text	Unique name of the entity-type
Description	Text	Description of the entity-type
Attributes	Text (SR)	List of references to documents of the type ATTRIBUTE

Document type RELATIONSHIP TYPE		
Attribute name	**Type**	**Description**
Definition	Text	Unique name of the relationship type
Description	Text	Description of the relationship type
Type	Value	'Binary' or 'is-a'
Member-1	Text (SR)	Reference to **one** document of the type ENTITY_TYPE
Member-2	Text (SR)	Reference to **one** (binary relationship type) or at least one (subentity types in a is-a relationship) document of the type ENTITY_TYPE
Cardinality-1	Value	'0,1', '1,1', '0,N', or '1,N' (null if is-a relationship type)
Cardinality-2	Value	'0,1', '1,1', '0,N', or '1,N' (null if is-a relationship type)
Attributes	Text (SR)	List of references to documents of the type ATTRIBUTE

Document type *ATTRIBUTE*		
Attribute name	Type	Description
Definition	Text	Unique name of the attribute
Description	Text	Description of the attribute

Document type *ORIGIN*		
Attribute name	Type	Description
Definition	Text	Unique name of the origin
Description	Text	Description of the origin
Type	Value	Origin type (for classification)

INDEX